CARRIAGES TO CARS

JAGUAR ROVER
MORRIS . AUSTIN . WOLSELEY
MORRIS COMMERCIAL
WIRES · ·SANDERS HITCHIN
PHONE HITCHIN 4436/7

RALPH E. SANDERS & SONS LᵀᴰRALPH E. SANDERS & SONS Lᵀᴰ

AUTOMOBILE DISTRIBUTORS

HITCHIN, HERTS

ENGLAND

MINIATURE IMPROVED LANDAUS & BROUGHAMS

ONE OF THE BEST SELECTIONS OF GOVERNESS CARS IN ENGLAND

CARRIAGES TO CARS

THE STORY OF COACHBUILDERS AND MOTOR DEALERS

RALPH E SANDERS & SONS
OF HITCHIN

Stephen Bradford-Best

A Hitchin Historical Society Publication

A Hitchin Historical Society Publication 2014
http://www.hitchinhistoricals.org.uk

© Copyright 2014 the Sanders Family, Stephen Bradford-Best, Hitchin Historical Society

ISBN: 978-0-9552411-9-2

Design, layout and photo enhancing: Barrie Dack and Associates 01462 834640
Printed by: Olive Press, The Green, Stotfold, Hitchin, Herts, SG5 4AN

Telephone 01462 733333

Cover illustration: The Walsworth Road garage, early 1920s.

Half title page: Letterhead of Ralph E Sanders & Sons Ltd, 1950s.

Frontispiece: An array of new carriages on display at the Walsworth Road workshops, 1900s.

Rear cover: The railway bridge Cambridge Road, Hitchin with Sanders' advertising hoarding, c. 1935.

The maps shown inside the front and back covers were provided by Hitchin Museum and used with the permission of Ordnance Survey.

CONTENTS

Introduction

My involvement with this project came about when a friend and neighbour, Stuart Sanders, grandson of Ralph Sanders, told me about the vast archive that he, and other members of his family, had stored in their homes concerning their family's former business as coachbuilders and motor dealers in Hitchin - Ralph E Sanders & Sons. The archive includes record books, a large collection of black and white photographs and other items from the time when Ralph founded his business in Royston, Hertfordshire, in 1875, through to the closure of the motor sales and repair business in Hitchin in 1979. Stuart was of the view that this valuable collection deserved a wider audience. Being a long standing member of Hitchin Historical Society I put forward this archive as the basis of a future Society publication; the Society has an enviable track record of publishing books on Hitchin's history. A positive response came from the Publication Team, headed by Society President Priscilla Douglas, and work commenced on examining the archive in more detail and carrying out further research to check the facts and to fill in the gaps. There were indeed gaps, particularly during the inter-war period, so I apologise if information for these years is a little sparse.

At some point, I can't remember when, it occurred to me that I had taken the lead with those initial enquiries and (shock!) somehow the task of actually writing the book had befallen upon me - and there was no going back. The 'team' held regular convivial meetings at the home of Stuart, and his wife Eileen, and I must thank them both for their hospitality - particularly the Bourbon biscuits! My subsequent enquiries included a search for surviving carriages and cars and I am pleased to say that eleven horse drawn carriages have (so far) been found. I enjoyed travelling to distant parts of the country to meet the proud owners but, sadly, I found no cars. The enthusiasm that I have encountered from those passionate about carriage driving has been heart-warming. Equally, the encouragement and assistance from renowned motoring journalists and historians, as well as from members of car clubs and many others interested in motor transport, has been invaluable. I will tell you more about those carriages that do survive, and possible reasons why no car survives, in the later section - 'Surviving Carriages and Cars'.

The consequence of all this effort on the part of many is a book containing a mixture of family genealogy, information about the carriages and bodies for motor vehicles built and repaired by Sanders, stories about their customers, their progression to the sale of motor cars and commercial vehicles, descriptions of the various buildings used by the company, reminiscences of family members, employees and others, together with an overview of the period covered. It is not a definitive work on coachbuilding (although you will learn much from reading Derek Wheeler's contribution) or even a complete history of the business but it is a record of the diverse information already in the family's possession combined with the facts, stories and anecdotes that the team and I have uncovered along the way, all very much with a leaning towards local history; after all, the publisher is the Hitchin Historical Society.

I confess that I have found this project, my very first, somewhat daunting; I have had short articles published before but never a whole book. However, I have strived to tell as full and as accurate a story as I am able, within the time available, and the resulting book has only been possible because of the considerable help and support that I have received from the Sanders family and very many others, combined with enormous encouragement from Priscilla Douglas. You will make your own judgment whether I have succeeded when you have read the book.

An advertisement for Triumph Cars, c.1930

Acknowledgements

The basis of this book is the archive held by Joanna Cooper (*née* Sanders), Robert Sanders and Stuart Sanders which includes many documentary and photographic items.

Like so many on Hitchin's history, this book would not have been possible without the comprehensive professional resources of Hitchin Library, Hitchin Museum and Hertfordshire Archives and Local Studies. Other organisations that helped were Bedfordshire & Luton Archives and Records Services, the British Schools Museum Hitchin, Hitchin Initiative, Hitchin Priory, Royston & District Museum & Art Gallery, Knebworth House Archives and Messrs John Shilcock.

In addition, I am most grateful to the following who have been very generous in providing help and information, for lending photographs and documents and for contributing memories:

Sybil Atkinson, Derry Aust, Robin Barraclough, Tiggy Bays-Griffiths, the late Tony Beadle, Sam Biles, Pamela Birch, Rutger Booy, Nick Boxall, Trevor Brockington, Debbie Charlesworth, Neville Chuck, Pamela Clark, Tom Clarke, Jenny Clune, Patrick Collins, Nicky Colman, Rory Cook, Nic Cooper, Tom Doig, Luc Eekhout, Reginald & Janet Fair, Andrew Fell-Gordon, Ian Ferguson, Clare Fleck, Susan Flood, Nick Georgano, James Gresham, Audrey Griffiths, Alexandra Gurr, Giles Guthrie, Raymond Hart, Oliver Heal, Michael Hill, David Hodges, David & Bridget Howlett, the late David Humphrey, Steve Jarman, Malcolm Jeal, Stephanie Kill, Paul Leggett, Marta Leskard, David Louch, Nigel Lutt, John Malamatenios, Iain Mansell, Louise Martin, Paula Martin, Vic & Betty Martin, Katharine Magruder, Michael Montgomery, Mark Morris, Bill Munro, Michael Palmer, Bob Prebble, Jonathan Rishton, David Rooney, Phil Rowe, Lydia Saul, Louis Sauter, Paul Savill, John Scorer, Ann Sylph, Neil Thomson, Neil Tuckett, Hans Veenenbos, Helen Walch, Barry and James Wheelwright, Richard Whitmore, Gwen Wigley, Jonathan Wilkins, Michael Worthington-Williams and Jonathan Wood.

The following have contributed additional written pieces:

David Biles MBE, Anthea Birch, the late John Brown, Geoff Clark, Joanna Cooper (*née* Sanders), Jenny Feaver, Denis & Nicola Fisher, Brian & Amanda Goodwin, Vivien Hampton, Pauline Humphries, Richard Lanni, Anna Mann, Malcolm Salter, Robert Sanders, Stuart Sanders, Doreen Tindale, Michael E Ware, Derek Wheeler MBE and William Woods.

Members of the wider Sanders family whose contribution of memories, photographs and artefacts have added considerably to our story.

The comprehensive photograph collections of the London Transport Museum and the Zoological Society of London.

A special thank you to the manager and staff of Kwik-Fit (GB) Ltd., Walsworth Road, Hitchin.

For his expertise with the maps included in the book we must thank John Lucas.

The artist whose original drawings, many specifically commissioned, appear in the book is Joanna Cooper (*née* Sanders).

To Bridget Howlett and Derek Wheeler MBE who expertly and swiftly proof-read the text.

I am indebted to Barrie Dack for his good-humoured guidance and for turning our text and pictures into an attractive and professional book.

If I have left anyone out, I hope that you will forgive me.

Finally, but by no means least, I would never have reached publication without the experience and tremendous help of the Hitchin Historical Society Publication Team – Pauline Humphries, Vicki Lockyer and lead by Priscilla Douglas.

A Sanders' advertisement for the Triumph Super Seven.

A whip socket drawn by Joanna Cooper née Sanders.

Foreword

By Michael E Ware
former Director of the National Motor Museum at Beaulieu

I was delighted when I heard that Stephen Bradford-Best was writing a history of the coachbuilders and motor dealers Ralph E Sanders and Sons of Hitchin. The country coachbuilder has long been overlooked by the motoring writer. The two encyclopaedias on my bookshelf which cover coachbuilding both leave out Ralph Sanders. This book puts this omission to rights. Like so many such firms it started in the horse-drawn days and gradually, as the motor car became more available, they built both types alongside each other. Unlike other firms of this sort Ralph E Sanders and Sons moved with the times and had a vibrant garage business, repairing and servicing cars and selling new and second hand ones. This book therefore crosses a number of boundaries. It is first and foremost about the local history of a major firm in the area and the people that ran it. It's about the art of coachbuilding, both of horse-drawn vehicles and of cars and commercial vehicles and also the nitty-gritty

of running a successful garage business. The background was in a wheelwright's shop in the 1840's and the firm continued in business until 1979. The book has a wonderful selection of illustrations of the company's products showing the wide range of horse-drawn vehicles - even an ostrich pulling a Governess cart at London Zoo! There are also a wide variety of pictures of many cars and commercial vehicle bodies. Though total production figures are not mentioned it is surprising that the author has only been able to trace 11 horse-drawn vehicles and no cars. Perhaps this book will tease out a few more.

In these days, when traditional book publishing appears to be in the doldrums, I do hope that this publication by Hitchin Historical Society will encourage other authors or local history societies to look at the possibility of producing books about coachbuilders and the motor trade in their areas.

Sanders' "Princess" Phaeton

Sanders' limousine body on a Rolls–Royce chassis

A Short History of Coachbuilding
by Derek Wheeler MBE

Derek Wheeler is a former Editor of the Hitchin Historical Society Journal. A trained craft teacher by profession, and latterly Technician to North Herts District Council Museums, he has restored both Victorian bicycles and 19th century fire engines. The subject of this book is close to his heart, since he once restored a pony phaeton for a local lady. Also, his grandfather, when applying for the position of chauffeur, was told that, if he could collect a car from Mr Sanders and successfully drive it through Hitchin on Market Day, he could have the job. That was his one and only driving test! The car was a big very old fashioned chain-driven FIAT and the date circa 1919.

We have been conversant with powered transport of varying sorts for well over a century and a half, yet there are still a few very elderly people who remember seeing their first aeroplane. There are those of us who, in childhood, were witness to the last vestiges of horse-drawn transport, since the family bread, beer, milk and vegetables were brought to the door by Dobbin, when petrol and motorised vehicles were still in short supply at the dawn of the new Elizabethan age. We imagine that three or four generations back our ancestors were incapacitated by total reliance on non-mechanised transport, yet there is no evidence to support this. They coped and they coped adequately. It was just a different way of getting around and planning for events.

In the 19th century, everybody from toddlers to great-grandparents was expected to walk and walk great distances. They walked to school, they walked to work. In rural communities they walked miles to catch a train, to do their shopping, to fetch the midwife, the doctor or the undertaker. This was the pattern of life.

The horse, the cart, the coach and the carriage (a lighter version of the former) were the prerogative of the moneyed classes – the farmers, the landowners and the professionals such as doctors and lawyers. This is not to say that the vast majority of the population were immobile. Because rural communities were so interdependent on each other, by the mid-19th century, when roads had improved, tradesmen invested in carts to distribute their wares out to the community and rural businesses such as poultry dealers, bakers, dairy farmers and nurserymen who brought their produce into the markets and towns. With the spread of the railways (and Britain had its greatest rail network before the First World War), parcel distribution, furniture removal and commodities such as pottery, glass, coal, ironmongery and perishable goods could easily be moved from station to customer by horse van. The horse was best on short journeys with frequent stops; it had the added advantages over a motor van that it stopped automatically at the door of regular customers and it contributed naturally to the success of local gardens! The bicycle introduced an increase of mobility but its carrying capacity was limited.

Because everybody knew everybody else, help with transport was always there. Enterprising carriers with their carts came into towns several times a week, transporting local produce and as an additional service, would pick up rural walkers. In Pigot's 1839 Directory in Hitchin Museum, there is a list of all the carriers who regularly came into Hitchin and a timetable for pickup arrangements, usually involving a village pub. This made sense, because the horse would need food and drink, as would the carrier and his passengers. This was the way the vast majority of the population travelled because it was comparatively cheap.

Seven road coaches came into Hitchin every week from all over the country until the railway was established. The most famous was the London coach, run for many generations by the Kershaw family, which stopped at Welwyn so that passengers could transfer there for a coach going north along the old A1 but this was a very expensive way of travelling. Each week a wagon travelled from Boston in Lincolnshire to London via Hitchin and obviously returned the same way. Sometimes people used the wagon service for their goods and furniture and walked behind it! It was cheaper that way and theoretically they could not get lost! The railway arrived at Hertford a few years before it reached Hitchin and an enterprising coaching inn offered a coach service between the two towns in the 1840s to meet the London train! Rail travel was a quarter of the cost of coach travel and many times faster. Kershaw's service was withdrawn three days after the first train reached Hitchin in August, 1850. This was the way forward. With the arrival of the railway, cabmen bought new vehicles to meet the trains and added a new business to the old inn yards which sorely felt the pinch when horse hiring fell into decline.

All of these horse-drawn vehicles, whether road coaches, wagons, dog carts for doctors, traps for governesses, spring carts for tradesmen and broughams for both gentry and cabmen, needed a vast infra-structure to keep them going. That service industry comprised the aforementioned inns for horse hire and stabling, suppliers of hay, and harness, saddlers, wheelwrights, blacksmiths and farriers. Add to this, coachbuilders, coachpainters, vets, and a very specialist industry, whose foremost exponents was a firm called Swaine Adeney, who made carriage whips. For us today to appreciate the range of this industry, one only has to visit the sale rooms of Thimbleby and Shorland in Reading during their specialist carriage auctions.

In earlier times farm carts and domestic carriages had

Kershaw's coach restored by Sanders for the Hitchin Pageant in 1951.
(John Scorer)

A drawing of a farm cart.

wooden bodies, wooden wheels and wooden axles, with bearing surfaces reinforced with iron. By Victoria's reign a great deal of sophistication had crept into the construction of these vehicles. Better turnpike roads, looked after by turnpike trusts and later by highways boards, allowed for lightness of construction which was kinder on horses, since it allowed them to travel further and faster without utter exhaustion and an early entry into the knackers' yard.

Improved metal technology meant that efficient springs could be coupled to iron or steel axles. Wheels became lighter and instead of the Roman practice of nailing hot iron plates, which were called strakes, over the joints in the rim sections (known as the felloes) of a wooden wheel, continuous iron hoops were shrunk on. These were known as tyres, since they tied the whole wheel together with tremendous compression forces. Wheelwrights could roll up their own hoops, fire weld the joints in a forge and shrink the tyres on to the wooden wheel on a special iron platform out in the yard. The rolling forces to which a wheel was subjected when in use meant that the tyres eventually came loose as the metal stretched. A temporary cure was to drive the cart into a stream to allow the wood to swell, but it only stopped the rattles until the sun came out!

By the 1840s much of the coachbuilding industry was mechanised. All the highly specialised and superbly designed ironwork had to be produced by smiths having access to forges, power hammers, rolling mills, lathes, presses and drilling machines. Whilst the local blacksmith could do skilled repairs, he could not necessarily reproduce in quantity and at speed the requisite fine metalwork of a carriage. Farm carts were his speciality.

This is where firms such as the subject of this book came in. There were specialist firms in the industrial Midlands and in London who could furnish coachbuilders such as Odell's and their successors, Sanders, in Hitchin, with purpose-built sundries, all made to pattern and gauge. Most of these fittings were made by machinery in the new steam age. There were Birmingham axle and spring smiths, nail and screw makers, brassworkers producing hinges and catches. There were leather workers producing upholstery and hoods and ironmongers by the score making carriage hardware. Foundries produced Warner hubs, which were a series of mortices in an iron cage, shrink-fitted on to wooden wheel centres, making the spokes rock-

Sanders' workshop showing carriages being constructed, including the
making of the wheels, with the forge in the background, c. 1908

The forge at Sanders' Walsworth Road premises, c.1908

solid at the root. There were also the specialist wheel builders who would import American hickory already steamed into a semi- circle, removing the need for a rim to be built up of 6, 7, or 8 felloes. They would also sell pre-cut traditional ash felloes or oak spokes and market mass-produced elm stocks for hubs.

There were also two new industries emerging; the first was producing solid rubber tyres, known as clinchers, for the carriage trade. These were sold in strips with a dovetail section at the base and a curved surface facing the road. A circle was bent up to the right size and cut to length. It only had to be persuaded into the steel rim (which was bought in, ready rolled to section) with a specially made machine with 4 feet of leverage on it and it was there for years. When the rubber wore down or got flats on it through over-enthusiastic use of the brake, it was back to the coachbuilder for re-tyring.

The second industry was that of carriage lamp building. Since every road vehicle needed two or three lamps for travel at night, specialist firms working in small premises employed men using hand presses, stamping out lamp blanks, plating mirrors and bevelling glasses, which were then assembled with soldering irons. Often these beautiful lamps hanging in pubs are the only evidence of a long-lost cottage industry. This was mass production long before Henry Ford got the credit!

Firms like Odell, Sanders and Cain in Hitchin, and their not-too-distant neighbour, Maythorn in Biggleswade and McMullen in Hertford, ordered the parts by post or telegram and waited for the Great Northern or Midland Railway to oblige. They then formed the components skilfully into a beautiful harmony of curves, chamfers and bevelled glass windows, with lightness and smooth flow of line being the essentials.

The woodwork was made and fitted in house. Ash frames were steamed and bent to shape and held under pressure over formers until dry. Joints were tenoned and dowelled, with wedges for added strength, since the animal glues of the day did not do well in damp conditions where strength was required. Sometimes iron plates or brass angles were screwed in place to assist wood in known areas of weakness. Panels were cut to patterns which were hanging on the walls. These wooden patterns could be worked up from design books which could be purchased from specialist suppliers. Panels often had to be bent in more than one direction because elegant curves showed class and distinction whilst contributing to

A sample of a coat of arms as painted on the door of a carriage

strength. This bending was done with hot water and weights, sometimes a gas jet being there to encourage the warping!

Door and body panels would be cut by hand with a bow saw or mechanically cut on a gas - or steam - engined bandsaw. Vibration, or 'drumming' as it was known, was often reduced by making a sandwich of hessian and checkerboard strips of scrap mahogany glued on to the surface. Upholstery then covered the surface. Leather hoods were made out of treated skins which were stretched whilst wet over the hood frames. When dry and thus formed, they were treated with copious amounts of leather dressing to make them waterproof and less liable to crack in sunny weather.

Coach panels were scraped, often with pieces of old window glass, and filled with plaster-based material until they were smooth. Then they were primed and rubbed and filled and primed and rubbed and undercoated. At the end of the process as many as 20 coats of paint and varnish had been applied by brush in an area shrouded in wet cloths, to avoid dust. Before the final varnishing, contrasting fine lines and gilding would be applied, with silver monograms added to door panels. Often brass beading was silver or nickel plated on expensive carriages. All coachbuilders had standard illustrated books so that they could entice their customers to buy a canoe landau, a governess cart, a ralli trap, a waggonette to take the village cricket team to matches or a smart Stanhope gig.

All ironwork, wrought to the very lightest of section and curved wherever possible, was fixed to the timber with so called 'coach bolts' which had long stems, large domed heads surmounting a squared section (which locked them into place both in wood and metal), and Whitworth threads - the standard thread of British engineering for more than a century. All nuts were square. One adjustable spanner fitted everything!

Whilst cart shafts, like felloes, could be purchased in sets, ready steamed and bent, coachbuilders often went out to purchase an odd tree or two for seasoning, because it had usefully-shaped limbs, thus obviating the need for purchase from a wholesaler. This also applied to material for making carriage poles for a two- or four-horsed vehicle.

The 19th century country coachbuilder gained more work from tradesmen such as butchers, bakers, cabmen and undertakers than he did from the grand families. The Delmé-Radcliffes, the Ransoms, the Seebohms and the Tukes from Hitchin, the Pollards of Highdown, the Goslings of Wellbury and the Lyttons of Knebworth used their carriages far less frequently than a country doctor or a grocer's boy with a dogcart. These were the people who notched up the miles and kept Odell's and their contemporaries busy with repairs. Whilst the aristocracy might go to Offords of London or Cockshoots of Manchester for their glass, D-fronted broughams, local tradesmen valued and could afford the work executed by craftsmen much closer to home.

These proud, skilled craftsmen, with hands used to shaping timber with draw-knife and spokeshave, saw no threat from the emerging motor car body. The same principles applied; save weight by cutting chamfers and give strength, coupled with beauty, by forming flowing curves. For many years to come their nostrils would still twitch to the fragrant scent of fine timber and linseed oil, albeit somewhat adulterated by the alien smell of petroleum spirit. Their skills had to adjust to building a little stronger in order to absorb the vibration of a heavy, slow-revving engine and to cope with speeds far greater than anything a horse could provide. This is the emergent world which saw the demise of Odell's and the heyday of Sanders of Hitchin.

A drawing of a motor car fitted with a limousine body.

*A Sanders' brass hub cap
fitted to carriage axles.*

The interior of Sanders' workshops with carriages and motor bodies under construction.

How it all began:
the Sanders Family and the Coachbuilding Trade

THE FAMILY TREE

John Sanders **m** Ann Hayden
b. 1783 · 1783 – 1855

William Nicholls **m** Elizabeth Baker
1815 – 1891 · 1812 – 1900

Thomas Nicholls 1845 – 1907	**William Nicholls** 1847 – 1896	**Martin Luther** 1849 – 1939	Ralph Erskine 1851 – 1933 **m** Rhoda Alice Wilkins 1851 – 1931		**Elizabeth Cromwell** 1853 – 1930

| Nora Ellen Jessie 1878 – 1968 m Robert Dredge | Ralph Francis Wilkins 1879 – 1949 m1 Alice Armstrong | Launcelot Vivian Erskine 1881 – 1970 m Daisy Custerson | Ida Kathleen Eliza 1882 – 1962 m Charles Dance | Rhoda Alice 1884 – 1947 | Marjorie Florence 1886 – 1968 | Doris Miriam 1888 – 1956 | Irene Mary 1889 – 1976 | William Nicholls 1891 – 1978 m Florence Jones | Jonathan Royston 1893 – 1980 m Elizabeth Goodman |

m2 Florence Upchurch

| Bernard 1905 – 1990 m May Wallace | Betty 1911 – 1988 m Wilfred Gatward | Mary 1920 – 1940 | Stuart 1936 m Eileen Hewitt | Peter 1920 – 2012 m Jean Deakin | Norah 1923 m Joe Russell | Bobby 1924 – 1949 | Molly 1927 – 1966 | Anne 1935 m Gerald Groves | John 1927 – 2006 m Jennifer Riley | Joanna 1931 m John Cooper |

Margaret Robert

Susan Diana

Nigel Ben Jon

Diana Ralph Bob

John Dudley Josephine

Lesley Tina Katharine Georgina Paul

Elizabeth Caroline Matthew

Nic Sally

My story starts with the wider history of the family whose descendant established the business, initially in Royston, Hertfordshire and later in Hitchin, the subject of my research. It is always fascinating to trace a particular family's long involvement with their livelihood. Ralph Erskine Sanders was born into a dynasty of craftsmen whose roots in the coachbuilding trade reached back to the early 19th century and branched sideways into local towns.

The 1841 census for Royston lists a **John Sanders**, Master Wheelwright, living in the High Street. He had been born in Essex in 1783. His elder son, named in the census as *'Nichols'*, aged 25, was described as a *'Coach maker'*. Nichols was actually **William Nicholls Sanders** (William senior), born in 1815, while his parents were living in the village of Hare Street, near Buntingford, as recorded on his marriage certificate of 1842. He took his bride, Eliza, the daughter of a Royston brazier, back to Buntingford, where he had set up what was to become a thriving coachbuilding yard in the High Street, close to the Market Hill area of the town. He and his wife had four sons and a daughter – **Thomas Nicholls Sanders,** born 1845, **William Nicholls Sanders** (William junior), born 1847, **Martin Luther Sanders,** born 1849, **Ralph Erskine Sanders**, born 1851 and **Elizabeth Cromwell Sanders**, born 1853. Three of these sons were to become coach makers in their own right. William senior

continued to work in the business until his death in 1891.

William senior's second son, William junior, helped to run the High Street, Buntingford operation, taking charge after his father's death. He was described in the census as a *'Carriage Maker and Timber Merchant'* and *'Employer'*. A timber merchant was a useful ancillary trade for a carriage maker as good judgment in the choice of timber was an important component in the construction of coach bodies and this expertise remained in the family, as described later. His eldest son Harry, aged 15, is listed as a *'Carriage Maker (employed).'* The census enumerator obviously did not recognise the term, and struck out *'Carriage'*, substituting *'Coach'* instead. William junior died, comparatively young, in 1896, at the age of 49 but Elizabeth, his widow, was obviously a capable person and took over the business. In 1901, she was described as a *'Carriage Builder'* and *'Employer'*, with Harry still her employee, but the timber-dealing side of the business appears to have ceased. Elizabeth died in 1904 and the Buntingford yard closed. Her son Harry stayed until the end and then moved, with his wife and family, to Brentford in Middlesex, where, in 1911, he was described as a *'Motor Cab Driver.'*

William senior's eldest son, Thomas, began his career under his father's watchful eyes. In 1861, he was living at home in Buntingford and already described as a *'Coach maker'*. Within ten years he had moved to Baldock and established himself in

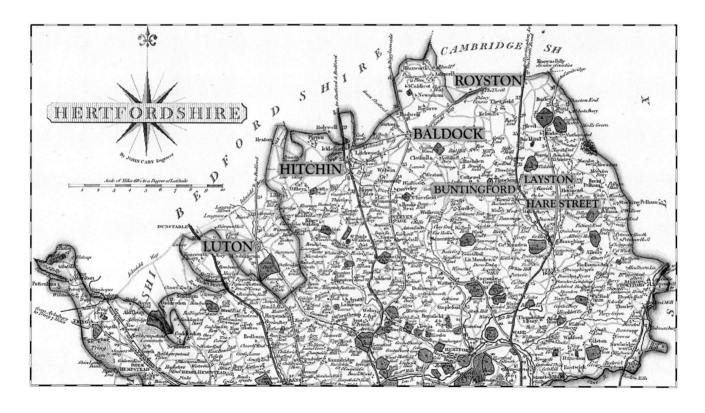

A map of north east Hertfordshire detailing those towns and villages associated with the Sanders' family in the 19th century

a yard of his own in Whitehorse Street. At the early age of 26, he was already a *'Master Coach Builder'* employing a man and a boy. He, and his wife Elizabeth, had three sons, all of whom followed him into separate branches of the trade. The 1891 census showed Thomas as a *'Coach Builder, employer'*, and his three sons were listed – Augustus as *'Coach Painter'*, Oliver – *'Coach Smith'* and another William Nicholls – *'Coach Body Maker.'* Thomas died at a young age, in 1907. Augustus (known as 'Gus'), the eldest son, continued in Whitehorse Street, although the business may have contracted later as the 1911 census showed no coachbuilding residents in his household, barring himself. Oliver had moved his skills to Hampstead in London by 1901. Thomas' youngest son, William Nicholls Sanders, possessed valuable transferable skills in the rapidly changing market place. By 1906 he was living in Luton, married with a young family, and described in 1911 as a *'Carriage Builder'*. He lived in Old Bedford Road in a domestic setting. The Vauxhall motor company had moved to Luton in 1905 and was by then well-established in the town, having become Vauxhall Motors in 1907. Did William work there?

The only one of William senior's four sons not to pursue a coach building career was the interestingly-named Martin Luther. He opted early for a railway career, becoming first a Clerk and eventually the Station Master at Harwich (St. Nicholas) on the Essex coast.

*A young Ralph Erskine Sanders
c. 1870s*

This leads to William senior's youngest son, Ralph Sanders, whose career forms the basis for this story. Ralph was born in 1851 and, initially, followed his brothers into the family business. From birth he was exposed to the minutiae of the trade and was to absorb this knowledge, along with much else, and apply it in later life. Not a great deal is known about his schooling. Census evidence records that, by the age of 16, his brother Thomas was involved in the family business as a coach maker, while his brother William, at the age of 14, was still at school. An assumption could be made that their father allowed the boys a good general education before he took them into the business as apprentices. Ralph was to demonstrate a real love of Shakespeare and Dickens throughout his life and could quote lengthy passages from the former, as well as extensive tracts from the Old Testament. He was particularly fond of Psalm 1 which makes reference to trees. There is a story that, occasionally, he would walk to London to attend Sunday Service. He was also a fine chess player.

At the age of 20, in 1871, Ralph was described as *'Smith'*, an essential contributor to the family business. Like elder brother Thomas, he soon branched out on his own account, choosing Royston, the town where his grandfather had been a Master Wheelwright, to establish his enterprise. An interesting history of Royston transport, published by the Royston Museum, traces his progress from a small yard in Kneesworth Street in 1875 to a larger one nearby in the following year. In 1877, Ralph married Rhoda Alice Wilkins, also born in 1851. By 1881, he was described as a *'Master Coach Builder'*, aged 30, living with his wife and two small children at number

A young Rhoda Alice Sanders c. 1870s

14 Kneesworth Street, next to the (then) Crown and Dolphin Inn, which was, incidentally, full of jockeys on census day! A census is only a snapshot in time, which perhaps explains the odd fact that one of the two infants resident in Ralph's household that day was a nephew from Baldock, while his own son, was conspicuously absent! Had he just been forgotten? Many census entries relating to the Sanders' family show evidence of great closeness between siblings and between generations, with a general tendency to offer employment opportunities, hospitality and support to relatives, as later examples confirm.

Ralph's attachment to Royston and its institutions remained pivotal to his life. Although he gradually transferred his commercial interests and energy to Hitchin, his family life was centred on the town. When his family outgrew the house in Kneesworth Street, they moved to a large house at York Villa in Great North Road, Royston. Ralph and Rhoda produced ten children, four of whom were to play important roles in the development of the family business. Later the family acquired a considerable property, Layston Park, on the outskirts of Royston, demolished in the 1970s to make way for a large residential development.

At age 47, Ralph expanded to Hitchin by buying an established coachbuilding business. The next chapter explains whose business he bought and later chapters tell how the business developed in Hitchin. Within the next decade, he also built new

York Villa, Great North Road, Royston, c. 1910

1933, has been a valuable source of further information. Obituaries do not always provide a balanced evaluation of character and achievement but, even with the hindsight of eighty years, his strikes one as both accurate and objective. Hine, who was a friend as well as Ralph's legal advisor, described him as *'a man of great tenacity and dynamic energy.'* Even allowing for some fine prose, for which Hine was famous (some say notorious), a real flavour of Ralph emerges, added to by many family anecdotes to complete the picture. Hine suggests that Ralph's greatest asset was probably his foresight, allied to his relentless energy. These characteristics were certainly in evidence in Hitchin demonstrated by the constant programme of expansion and refinement of operations which took place in order to cater for the most advanced bodybuilding and motoring requirements. None of these tasks could have been accomplished without, as Hine noted, Ralph's *'relentless activity'*. He never sat still, rose at dawn, and adhered to his maxim: *'the right time to begin the day's business is to begin before the day!'* He is said to have referred to Christmas as *'enforced idleness!'*

showrooms in Royston, although, by the end of the First World War, the company had ceased to operate in Royston and all its energies were focussed on Hitchin.

By the standards of any age, Ralph was a remarkable man. His obituary, written by noted local historian, Reginald Hine, and published in the *Herts & Cambs Reporter* dated 1st September

Some accused Ralph of impulsive action but, in the commercial realm at least, there was usually a basis of careful evaluation. It was alleged that his eldest son Francis, known as Frank, born in 1879, was given twenty minutes notice before being taken to Hitchin to be apprenticed to Walter Odell. But

Layston Park, Ralph's final home on the outskirts of Royston, c.1920s

an authority on English timber. Apparently *'he could cube-up and value standing trees at a glance.'* His youth spent in the Buntingford yard must have sown the seeds of this knowledge. It also gave him a countryman's love and appreciation of the land. One of the estates he purchased was Wood Hall in Arkesden, near Saffron Walden, which has an impressive house dating back to 1652, built on the site of a Saxon manor recorded in the

evidence from Sanders' Day Books confirms that Ralph had dealt with the firm for several years, must have known the business and trusted the family. The 1901 census shows Frank, then aged 21, as *'coach builder, worker'* occupying a house next door to Herbert Odell, aged 39, *'Coach Builder's Manager'* and his family in Bridge Street, Hitchin. Two of his young sisters, aged 17 and 15, were staying with Frank on the date of the census. I believe that this is a proof of both a structured commercial transaction and one of trust between the two households.

Examples of instant decisions to purchase swathes of woodland were also cited as recklessness but Ralph had become

Sanders' new showroom built in Kneesworth Street, Royston in 1907

Domesday Book. According to local historians it was here the first Congregationalists signed their covenant on 22nd December 1682. However, Ralph only wanted the estate for the timber; the house was let to tenant farmers and later sold.

Of course, a man as successful as Ralph might not have been easy to deal with. Hine succinctly commented *'One felt, and he knew, that he was pre-destined to succeed …'* However, not all his judgments were sound as Hine continued *'Off his beaten track, his second thoughts, not his first, were best.'* A good example of this is the sad tale of the Cottered tree. Ralph, having acquired the Lordship of the Manor of Cottered, proceeded to fell the ancient and much revered elm tree, nicknamed "Bumpy" by the residents, which had stood on the village green; presumably he wanted the wood. Outrage ensued but Ralph neatly side-stepped the Courts by making the residents a voluntary gift of the village green, in perpetuity – a masterful solution. Hine noted that if confronted with a legal point that went against his interests, Ralph declared the law *'an ass, and a*

"Bumpy'" the elm tree in the village of Cottered cherished for decades by the local residents.
(Neville Chuck)

rogue into the bargain.' He would brook no opposition.

It would be assumed that a man of Ralph's standing, with his increasing affluence, would have adopted habits of dress and behaviour in keeping with his new status. However he remained a man of very few pretentions. He loved work for its own sake and did not measure success in monetary terms. Indeed he had a casual attitude to personal spending and anecdotally kept his money in a shallow waistcoat pocket rather than a wallet. Famously, it was said, he once found himself on a train to Paris with only three shillings in his possession. He appears to have

"Bumpy'"after being felled by Ralph's men and which resulted in his donation of the village green to appease the aggrieved residents of Cottered.
(Neville Chuck)

Ralph Sanders c.1930

survived! This thinking extended to his personal appearance. Hine described Ralph as *'shabby and unconcerned.'* He apparently never wore shoes but favoured old-fashioned buttoned boots, only fastening the top one. When asked in jest if he would care to donate his favourite overcoat to the proposed Hitchin Museum, he smartly replied that he would consider it in his Will but meanwhile he found it comfortable.

With such a powerful and forthright character, Ralph not only enjoyed the position of Chairman and Joint Governing Director of his own company, from 1909 until his death in 1933, but played a prominent role in his home town of Royston. He was an original member of the Urban District Council, and later became a Director of Royston Gas Company.

Stuart Sanders, a lifelong railway enthusiast, recalls his favourite tale of his grandfather. 'This story was told to me by my father after my grandfather's death. My grandfather travelled much in pursuit of the business with the object of obtaining the best components and materials for the bodywork of the cars built by the company. There was an occasion in the 1920s when, returning by train from the north of England, the train was running late and the booked connection at Hitchin for the last train of the day to Royston was not held but released leaving my grandfather stranded at Hitchin. He was not well pleased to say the least and caused a considerable rumpus to such an extent that the station master arranged a special train to get him home. I heard about this incident again when, in about 1959, as an articled clerk to a firm of accountants, I was engaged in preparing accounts for an electrical firm, Pepper & Haywood, in Royston. This firm employed an elderly part-time book keeper and on being introduced to him he informed me of the time when some 30 or more years before he was employed by the railway at Royston station and he remembered a telegraph from Hitchin advising of the special train due to arrive later that evening. At first he was surprised and could not believe this was happening but sure enough the train duly arrived and out got my grandfather! The elderly book keeper clearly remembered this well and his account was in accord with what my father had told me.'

Another family story about Ralph refers to his longstanding battle with Royston Golf Club who had taken over part of Therfield Heath (sometimes referred to as Royston Heath) and, as well as laying out a golf course, had built a club house on what, in Ralph's view, was common land. A fence had been erected around the club house and Ralph simply removed it which led to some acrimonious exchanges. The club responded by making him an honorary member but to emphasise his strongly held view Ralph arranged for members of his staff to occupy every part of the club buildings, including the toilets, on one day each year. He also 'stored' a pile of wood on one of greens on the basis that,

as the land was a common, he was equally entitled to use it but, after repeated demands from the club, he eventually agreed to remove it. The long term effect of Ralph's actions is that much of the heathland is, to this day, still a public open space.

Ralph remained active and involved in his business life into old age. He retained his physical and intellectual vigour almost until his death, at the age of 82, in 1933. His wife of 53 years, Rhoda, predeceased him by two years and he was survived by all ten of his children; four of them engaged in the family business Ralph Francis, Launcelot Vivian Erskine, Doris Miriam and Jonathan Royston.

In this chapter I have referred to Ralph Erskine Sanders simply as Ralph. In the remaining chapters I will refer to the main family members concerned with the Hitchin business using the names by which they were better known: Ralph's four children, **Ralph Francis Wilkins Sanders** as Frank, **Launcelot Vivian Erskine Sanders** as Lance, **Doris Miriam Sanders** as Doris, **Jonathan Royston Sanders** as Roy and, later, his grandson, **Ralph Bernard Sanders** as Bernard. To remind you where they fit into the family I recommend that you peruse the Family Tree, beautifully hand drawn by Joanna Cooper (*née* Sanders), at the beginning of this chapter – their names are highlighted in blue. I will refer to the Hitchin business as Sanders or Sanders Garage. There is also a map of north east Hertfordshire on page 4 which may assist you with the location of the places referred to in this chapter.

The family gathering to mark Ralph & Rhoda's Golden Wedding Anniversary in 1927. Top row: **Jonathan Royston Sanders, Doris Miriam Sanders,** *Charles Dance, Irene Mary Sanders, Augustus Sanders,* **Alice Sanders,** *Robert Dredge and William Nicholls Sanders. Middle row seated: Rhoda Alice Sanders,* **Launcelot Vivian Erskine Sanders,** *Nora Ellen Jessie Dredge (née Sanders),* **Ralph Erskine Sanders,** *Rhoda Alice Wilkins Sanders, Ralph Francis Wilkins Sanders, Ida Kathleen Eliza Dance (née Sanders), Florence Sanders and her baby daughter Mollie Sanders. Seated on rugs:* **Ralph Bernard Sanders,** *Ralph Peter Sanders, Irene Mary Sanders, Robert Sanders, Norah Russell (née Sanders) and Marjorie Sanders.*
(The names of the family members actively involved in the business are in bold.)

The Odells:
their businesses and occupations

When Ralph decided to expand his business, he chose to buy an established firm of coachbuilders in Hitchin, rather than starting from scratch, although he was shrewd enough to purchase some additional land near the railway station with further expansion already in mind. But whose business did he choose to buy?

Hitchin, which has origins in Saxon times, is an historic market town in the north of Hertfordshire, about 35 miles north of London. In 1900 it had a population of around 9,000. It

Bridge Street in the 19th century showing the distinctive 'Dutch gable' style outline of the Odell Carriage Repository on the right

has good transport links being a junction station on the Great Northern Railway and close to the Great North Road. There were many coachbuilders competing for business in the town during the late nineteenth century, like John and William Cain, Alfred Rogers and William and Joseph Richardson. One of those coachbuilders, and possibly one of the larger ones, was the well established firm of Odell, lastly run by Walter Odell. They had extensive premises at numbers 13 and 15 Bridge Street close to Priory Park; number 13, to the left of the Boot public house, was a two-storey house having a room over a gated entrance to a workshop, warehouse and yard behind and number 15 comprised a building known as *'The Carriage Repository'*, or *'The Carriage Works'*, to the right of the Boot. (Numbers 13 and 15 were renumbered 26 and 28 around 1930).

The name Odell is not uncommon in Hitchin but I am fairly confident that Walter's grandfather was John Odell, born in 1803, and that he lived in the Collins Town area of the town, close to the Highlander public house, where he was described as a *'Master Blacksmith'*. He was married with four sons and later moved to Old Park Road, possibly on his retirement. His eldest son, Robert, born in 1828, followed in his father's footsteps and operated as a blacksmith in Parcell's Yard, off Bucklersbury. He was married with five sons – Stephen, Walter, John, Frank and Herbert – and, by 1861, was described as a *'Farrier'* living

Odell's Carriage Repository at 15 Bridge Street and shop to the right at number 16, c.1890

in Bridge Street *'next to the pub.'* I believe that he exercised his skills further up Bridge Street, near the Plough public house, different premises to those occupied by our branch of the family who operated, also with a forge, next door to the Boot public house at number 14 (later renumbered 27). This is born out by the census entry for 1881 when Robert was described as a *'Smith*

and Coach Maker' living at number 19 Bridge Street, close to the afore mentioned Plough public house and by the River Hiz, together with their middle son, John.

However, it is Robert's second son, Walter, born in 1853, who is relevant to my research. In the book *'Old Hitchin'* by Alan Fleck and Helen Poole (1976) there is a lovely story about Walter. It tells that the Odell family had been Baptists for generations and that, one day, Walter's father discovered him reading a harmless boys' magazine and, as a consequence, cast him out of the house. Some years later Robert discovered that Walter was serving in the Royal Horse Artillery and, suitably contrite, bought him out of the army. Walter then started a coachbuilding business in Portmill Lane and, in 1881, was living in Walsworth Road. Later, certainly by 1891, Walter was recorded as living at 13 Bridge Street, with his wife and daughter, and had joined the business run by his brothers John, Frank, and later Herbert, although, by 1891, Frank had left the area to pursue a career in hotel management in the West Country. Just for the record, the eldest son, Stephen, was also a smith and farrier operating from Old Park Road and later in Kent Place. He was married and had two sons, the elder, another Robert, also a blacksmith.

But what did the business comprise? Among the services that Walter offered to his clients, recorded in a Memorandum of the 1890s, was *Any Carriage Let on Hire with Option to*

Purchase.' He also supplied *'Lamps, Rubber Mats, Rubber Brake Blocks and all Carriage Fittings'* and *'Waterproof Driving Aprons.'* It is likely that *'Special Facilities for Storing Carriages at Low Rates'* would have appealed to those residents of the town who did not have sufficient space to keep them at home. *'Estimates Furnished Free for Repairs & New Work of Every Description'* was directed to those needing new transport or the repair of their existing vehicles. Also, like the modern motor car, there would also have been a need for *'Dunlop Pneumatic and all kinds of Rubber Carriage Tyres'* to replace those that were worn out. The workforce of skilled craftsmen made a variety of small carriages in their workshops tailor-made to the clients' requirement which would also have

Receipt dated 31st March 1897, signed by Walter Odell, for the premium paid by Ralph on his son, Frank's, apprenticeship

included farm vehicles as well as those for carrying passengers, even a horse drawn hearse for those needing transport to a higher place.

With his business in Royston firmly established for almost a quarter of a century it was not surprising that Ralph decided to expand and to seek new clients in a different location. He could have bought a site in another town and built new premises, although he probably avoided Baldock the preserve of his brother Thomas, while Buntingford was spoken for already. However, this option would have been time consuming and mean extra work convincing potential local customers that he could provide, at the very least, as good a product as established coachbuilders in the town, let alone convince them that he could do a better job at a more competitive price. A sounder alternative would be to purchase an existing business which would provide premises, a skilled workforce and, probably, a loyal client base. He would then be able to 'hit the road running' and concentrate on using his acknowledged business skills to further build up and expand his business. But should he buy a well respected business with a good reputation or simply one at a good price? He was aware of Walter's firm in Hitchin and presumably considered his products to be of good quality as he had purchased new and second-hand carriages from him, recorded in Sanders Day Books for 1897 and 1898 – e.g. *'5ᵗʰ November 1896 Paid Odell £35 for S/H*

Receipt dated 1st July 1899, signed by Walter Odell, for a part payment made by Ralph Sanders for his business.

Landau' and '*13th November 1897 Paid Odell £65 for six Governess Cars'*. Walter may also have been a personal friend. In addition, on 31st March 1897, Ralph had purchased an apprenticeship for his son Frank from Walter for the sum of £12 and Frank had started work the same year, further paving the way for Ralph to establish a Hitchin operation. In any event, for whatever reason, Ralph made an agreement with Walter to buy his business.

The Memorandum of Sale, dated 1st July 1899, states that Walter agreed '*to sell the goodwill of the business of a coachbuilder now carried on by him … together with the whole of the stock, machinery and utensils in trade*' to Ralph '*for the sum of eight hundred pounds sterling'*. The Memorandum also records that the amount was

to be paid by instalments. The first instalment of £150.00 was paid by two cheques, one of £50.00 on 28th June 1899 and a further cheque of £100.00 on 1st July 1899. I think that it can safely be assumed that Ralph paid the balance in accordance with the agreement. In addition to the '*goodwill of the business*' Ralph also purchased the various buildings owned and used by Walter fronting Bridge Street for the sum of £1,750.0s.0d., including the distinctive Carriage Repository together with the cottages, outbuildings and gardens behind and all rights of way, as detailed in the original Indenture dated 1st February 1902. The Indenture, signed by Walter, also states that he '*will not at any time hereafter…..be engaged in the business of a Coachbuilder within a radius of six miles from Bridge Street, Hitchin'*.

What caused Walter Odell to sell his business is not known; his retirement is a possibility. However, he was not idle as, in 1901, he is described as a second-hand book seller living at 16A Bridge Street. He later went on to establish a second-hand bookshop and lending library at 20 Churchyard in Hitchin. In his book, *Confessions of an Uncommon Attorney*, Reginald Hine said of Walter that he '*knew next to nothing about books*' and that he sold them simply '*by shape*', not by their content, with the books littered on the floor, amongst dust and debris, always threepence. Despite this change of occupation his advice about the carriage trade, and his expert valuations, were continued to be sought and he even wrote regularly on this subject for *Exchange and Mart*.

The sale to Sanders ended the Odell family's involvement with the Bridge Street properties; well not quite, as Herbert continued as manager of the coachworks for a while, probably until Frank completed his apprenticeship, and, as mentioned above, Walter continued to live at number 16A Bridge Street (later re-numbered 29) until his move to Churchyard.

The Sanders' carriage store, c.1908

18

The Business Expands:
the lead up to the First World War

In this chapter I will tell you about the expansion of the business in the lead up to the First World War. It is the largest chapter by far, partly because the business activities expanded rapidly during the first decade of the twentieth century but also because most of the Sanders' archive material is for this period, including information about the increasing number of buildings on the two sites and the wide range of products and services that the company offered to its customers.

I will start with a description of the buildings – both those purchased from Walter Odell in Bridge Street, including the replacement of one of them, and the several that were built specifically for Sanders on a greenfield site in Walsworth Road.

Bridge Street

Ralph had purchased from Walter Odell the goodwill of his business on 1st July 1899 and the range of buildings he owned in Bridge Street on 1st February 1902. A plan inside the front cover of this book shows what I believe to be the extent of Sanders' premises when their business expanded to Hitchin. The buildings comprised numbers 13, 15 and 16 Bridge Street (later re-numbered 26, 28 and 29), together with the yards, cottages and gardens behind, and were where Frank continued to work, eventually to become manager once he had completed his apprenticeship, although, by around 1902, the main workshops

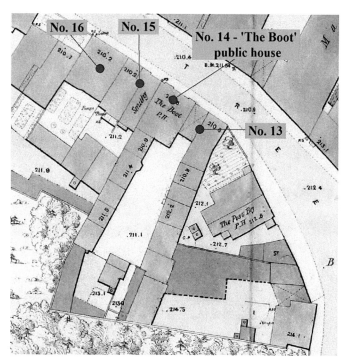

Plan of Bridge Street showing the street numbers of the properties owned by Sanders in the early years of the 20th century.

had moved to new purpose-built buildings in Walsworth Road. I believe that, before the move, the Bridge Street business was initially managed by Herbert Odell, presumably as a Sanders employee, and who, according to the 1901 census, was living with his wife and three children further up the road at 11A Bridge Street, the former Post Boy public house and now part of the well known second hand bookshop founded by Eric T Moore. A useful source of information from The National Archives has become available online recently – known as the 'Hitchin 1913 Project' – resulting from proposed government legislation relating to a property tax, included in the Finance (1909-10) Act 1910, which was not, after all, put into effect. The survey associated with the legislation records details of all properties in each town throughout the country, together with their value, and was carried out between 1910 and 1915. It has proved a useful guide.

Number 13 Bridge Street was a small timber framed and rendered two storey house with a small garden on the left hand side. On the ground floor was a door to the street and a single window at the side. On the first floor there were two windows, the second in a part of the building which extended over an archway attaching it to the next door Boot, later Royal Oak, public house at number 14. Sanders bought number 14, in 1923, from Cosmo Brett and others, believed to be directors of Fordham's Brewery of Ashwell, by which date the public house had closed. It was then let to tenants, including Arthur Randolph and later D & E Petty (see 'Bus Garage' in the next chapter). The archway gave access to a yard at the rear where there were warehouses behind the house. I know that there was a diamond shaped plaque between the two first floor windows

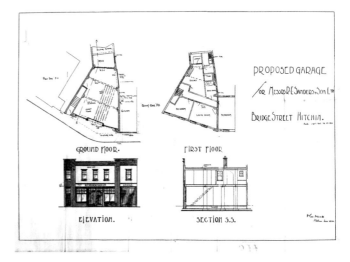

The plan, dated 1914, of the new showroom built for Sanders at 13 Bridge Street. (Hitchin Museum)

The new Sanders' premises at No. 13 Bridge Street soon after it was built with period displays and advertisements in the windows and a car parked under the archway.

at the front, probably recording a date, but I do not have details of the inscription. According to the 1901 census the house was occupied by Frank Sanders and Kelly's Directory for 1902 records Sanders also being in occupation of the business premises but they had left by 1905. Hitchin Urban District Council (HUDC) rate books record that the garage was subsequently occupied by Slater & Co (sometimes referred to as Slater, Batty & Co) and

then Sanders again from 1915. Slater, Batty & Co also occupied premises on the opposite side of the road at number 9A, part of the old malting building. Around this time the original building at number 13, purchased from Walter Odell, was demolished to make way for a new purpose built car showroom for Sanders, with a flat above. I am unsure of the exact date that it was built but it was probably 1915, when HUDC rate books record that Sanders had returned to this address, and certainly by 1920. The plans for the new building, dated 1914 and signed by local builders John Willmott & Sons, detail a much larger structure on the site, taking in the garden at the side of the old house to fill the full frontage of the plot and reaching back to cover some of the land previously occupied by warehouses. There is an atmospheric photograph of the new building with 'R E Sanders & Sons Ltd' sign written over the door after they had moved in. The two bedroom flat, on the first floor, had independent access from the street and comprised a kitchen, a scullery and walk-in larder, two bedrooms and a bathroom. It was occupied by Brown in 1918 and Bottoms in 1919. This building was sold by Sanders in the 1930s; its later occupants are described in the final chapter.

Number 15 Bridge Street appears to be a purpose-built carriage works but I do not know whether it was built for the Odell's or an earlier occupant. However, nineteenth century photographs show it emblazoned with the title 'Odell's Carriage vacated the building by at least 1905, possibly earlier, and let it to a tenant. The 'Hitchin 1913 Project' gives details about this property, recorded on 1st December 1914, when it was described as showroom, workshop, stores and gardens and that, from 29th September 1905, Sanders had granted a lease to Herbert Frost, 'antiquarian dealer', for a term of 5 years at a rent of £50 per annum. It appears that Frost displayed furniture for sale in the showroom with an adjoining workshop. HUDC rate books record that it continued to be occupied by Frost from around 1911 to 1921. He was then replaced by Arthur Randolph, also described as a furniture dealer. Both Frost and Randolph also occupied other buildings in Bridge Street.

The adjoining property in Bridge Street appears to have been divided into two parts number 16 was a shop, Sanders taking over occupation from Odell on the purchase of the business, and number 16A was living accommodation where Walter had lived with his family in the very early years of Sanders ownership and where he commenced his new occupation as a second hand book seller. HUDC rate books record that, together with number 15, it was later occupied by Herbert Frost, certainly from around 1911 to 1919. Both numbers 15 and 16 were rebuilt in the 1960s as mentioned in the final chapter.

The Bridge Street premises soon after purchase – the old carriage works at No. 15 on the left, and the shop and house, at No. 16, on the right, both with Sanders name over the doors.

Repository' in large letters on the Dutch gable style frontage. At ground floor level there were wide opening doors to allow the movement of carriages and full height double doors on the first floor, below the gable, to enable carriages to be stored aloft, presumably raised by an external hoist. Sanders must have

Walsworth Road

Having established his business in Bridge Street Ralph soon sought to expand it – but where? He required a greenfield site close to the town centre which would give him the space to build large new workshops to manufacture carriages and showrooms to display them, mindful of the coming motor age. Around 1900, he chose a site in Walsworth Road at the opposite end of the town from Bridge Street and close to the railway station. The site was part of land described on the 1844 Tithe Map as *'arable'*, called *'Dunche's Field'* and then owned by William Wilshere MP. The original plot, numbered 857, fronted both Walsworth and Nightingale Roads, the two roads having forked, heading west, as you enter the town from Walsworth, to form two sides of a triangle. However, the part of this plot that he purchased from Mr George Harding Innes was a narrower triangulation fronting Walsworth Road only. According to the 'Hitchin 1913 Project'

The range of three buildings built by Sanders in Walsworth Road in the first decade of the 20th century - the new car showroom on the left, the workshops and stores on the right, which included the lift to the first floor, and the original timber framed carriage workshop in the centre. Note the first floor 'bridge' which allowed vehicles to traverse the full length of the buildings at first floor level. The building on the left, and the central building, survive in the care of Kwik-Fit (GB) Ltd.

The Walsworth Road premises around 1910 looking north east towards the railway station.

£823 *'to be paid monthly to the extent of 80% of the work completed, the remaining 20% to be paid at the expiration of 1 month after the completion of [the] contract'.* The gable end, above two wide glazed windows and two further glazed windows to the left, are visible in an early photograph and can still be seen today at first floor level, although this building is now clad in corrugated metal. To the left of this building, a brick built showroom was

the sum of £820 was paid. The remaining part of the plot was developed by Innes and became the manufacturing base for his agricultural and general engineering company known as G H Innes & Co and later, after being joined by George Walter King, Geo. W King Ltd. Ralph conveyed the land he purchased to the new limited liability company, Ralph E Sanders & Sons Ltd, when it was formed in 1909. Soon after purchase, around 1902, he instructed a builder, Jacklin & Co of Royston who Ralph had employed previously, to construct a large timber framed two storey workshop set back from the road at an estimated cost of

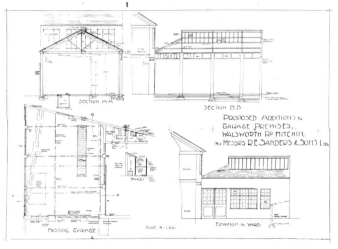

The plan of the rear extension to the Walsworth Road workshop c.1912. (Hitchin Museum)

25

constructed around 1906, claimed to be one of the first purpose-built car showrooms in the country, consisting of two blocks of two storeys, the one closest to the road having a steel girder roof construction and the rear block having timber trusses, both under tiled roofs. The street frontage had an elegant facade comprising three arches formed of very thin bricks with glazed openings, the centre one with double sliding doors giving access to the ground floor showroom which had a central staircase to the first floor; all since removed. However, still in place on the first floor is an oriel window in the centre at the front and an ornate semi circular window on the right hand side. A car inspection lift was later installed in front of this building.

The next building to be constructed, built around 1909 by John Willmott & Sons for £1,056.5s.7d., was at the extreme right hand side of the site and comprised a large two storey brick building of two bays with a tiled pitched roof over each bay, the right hand bay having a wide entrance on the ground floor. Once inside this entrance you came to a lift operated by a large flywheel and ropes, big enough to convey a carriage or car from the ground to the first floor. Both Stuart and Robert Sanders have childhood memories of being admonished by their respective fathers for swinging on these ropes. At a much later date the lift was removed and replaced by a curving ramp necessary to cope with the increased weight of motorised vehicles. It was on the

first floor of this building that the locally famous Kershaw's coach was stored until the 1960s. The left hand bay, on the ground floor, was occupied by the stores and the street window usually had displays of motoring equipment and accessories, often sponsored by a supplier, like Exide batteries. A first floor enclosed bridge (like the Bridge of Sighs), big enough to accommodate a carriage, connected this building with the original timber framed building

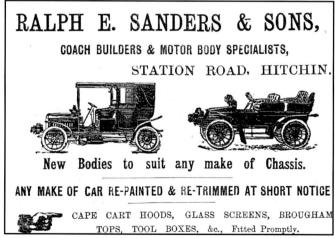

A Sanders' advertisement in a local trade directory dated 1906. Station Road later was named Walsworth Road

The enlarged Walsworth Road premises around 1919, with the new car showroom on the left and workshops and stores on the right, including a lift to convey vehicles to the first floor

in the centre of the site and, indeed, to the first floor of the car showroom on the extreme left enabling a carriage or motor car to cross from the far right to the far left of the buildings at first floor level. Against this building, to its left, was an open-sided pitched roof structure with decorative twin columns at the front, and a total of five columns at the left hand side to support the roof, which provided cover for cars on sale outside. On the ground floor, at the rear of the main building, there was a forge which remained in use until the 1950s.

The Sanders Property Register has been a useful source of information about the various properties already mentioned,

Sanders' advertising themselves as both cycle and motor manufacturers, c.1905.

although dates are sometimes ambiguous. It also gives information about further properties owned by Sanders. They include buildings in Tilehouse Street, Hitchin, bought from W & S Lucas Ltd, the brewers, in 1924. These buildings are probably the ones described in the HUDC rate books which record that Sanders owned 'buildings, yard and premises' at 7-10 Tilehouse Street. Around 1923, Ralph established a motor business in Castle Street and Pegs Lane, Hertford, the premises later leased to Eastern Automobiles Ltd and sold to Chaseside Motors Ltd of Enfield, at the time of Ralph's death, in 1933. In addition, also around 1923, Henlow Garage Ltd was established on land at Lower Stondon, Bedfordshire, but the date of its sale is unknown.

Company incorporation/shareholders

As the business grew, Ralph must have decided that it would be better administered as a limited liability company. The company was incorporated on 28th September 1909 as 'Ralph E Sanders & Sons Ltd' with capital of £15,000 divided into 7,500 Preference shares and 7,500 Ordinary Shares, all of £1 each. Ralph and Frank were the original shareholders.

Rhoda Alice Sanders (Ralph's wife) and Lance both first received shares on 12th March 1910 – Rhoda 500 shares and Lance 300. In December 1919 further members of the family, all being Ralph's children, received shares for the first time –

William Nicholls Sanders (Bill), Royston Jonathan Sanders (Roy), Doris Miriam Sanders, Irene May Sanders, Margery Florence Sanders, Norah Ellen Jessie Dredge (née Sanders) and Ida Kathleen Eliza Dance (née Sanders) – Bill and Roy 300 each and Ralph's daughters 100 each. They all added to their shareholdings over the years.

Coat of Arms

The company displayed an impressive coat of arms on its letterheads, invoices and other documents. Its origins are a mystery to present family members and it is unlikely to be authentic and registered with the College of Arms, the official heraldic authority for England. The choice of what appears to be the sun god Apollo (or Phoebus) in his golden chariot (above the helmet) is apt for someone in the carriage trade. The Latin motto – 'Surgit Post Nubila Phoebus' – loosely translated means 'after

Sanders' coat of arms – possibly inherited from Odell but unlikely to have been approved by the College of Arms

clouds, sunshine'. However, whilst looking through invoices headed with Odell's name, I noticed that they also displayed a very similar coat of arms with a variation on the motto – 'Surgit Nubila Phoebus'. An innocent means of raising the status of the company.

Personnel

During this period, Frank, having served his apprenticeship with Walter Odell, managed the business on his father's behalf. He was joined by Lance whose expertise was carpentry and upholstery.

I know far less about the people who worked for Sanders during this early period, except, of course, for the family members. I mentioned before that, according to the 1901 census, Herbert Odell was described as the manager of the business. However, by the 1911 census, he was living in Bishop Auckland, County Durham with his wife and five children, having previously lived in Shefford, Bedfordshire and Hendon, in north London, after leaving Hitchin.

Names of other employees are included in an Account Book which appears to span the ownership of the business by both Walter Odell and Ralph and which gives some information about the wages they were paid around the turn of the century. One employee, named Sharpe, probably a child or a trainee, is first recorded in 1896 when he received 1/6d per week, increasing to 2/– per week in early 1897, 3/– per week in late 1897, 7/– per week by early 1901 and, by late 1901, increasing to 8/– per week. Another, Wilson, possibly a foreman, received £1:18s:0d per week in 1901. Lastly, Frank, who had started working for Walter Odell by at least March 1897, received payment of 15/– per week in 1901. There is a carbon copy of a letter written by Frank, on 4th November 1908, asking a Mr Butterfield to prepare an apprentice's indenture for a '*lad*' named '*Levi Hawkins, aged 14 on the 30th September last, son of Charles Hawkins*'. The apprenticeship was to be for a period of five years with a premium of £10. His pay was to increase annually from 3/-, to 4/-, 5/-, 6/6 and finally to 8/– per week during the period of the apprenticeship. Under the heading 'Name of Charity' is recorded Gyvor's of local village St Ippolyts, implying that the premium was paid by this charity on Levi's behalf. Research has revealed that William Gyvor owned land in St Ippolyts and that, by his will proved on the 20th April 1654, he gave a perpetual annuity of £4.0s.0d from a rent charge on 3 acres of land at Pound Close, St Ippolyts to apprentice a poor boy of that parish. Levi was the fortunate beneficiary in 1908.

Also in the Sanders' archive are the original indentures for several apprenticeships which commenced in this period, including John Newland from 1899, Charles Holland from 1900, Christopher Norris from 1902, Frederick John Bysouth from 1903, Charles Anthony Willey and Frederick Bertram Waldock from 1914, Frank Jardin Sullivan from 1916 and Ivan Dearmer from 1918. How long they remained with the company and what skills they learned is not recorded.

In a letter, dated 13th August 1942, Frank refers to a Day Book, probably for 1911, and says that the entries are either in his own hand or those of Mr R G Chandler '*who is now one of the Governing Directors of Henly's. He came with us shortly after leaving school and continued here right through the last war*'. Henly's was a large company who owned a nationwide chain of motor dealerships.

Timber

I know that Ralph toured the south east of England looking for farms to buy, or more particularly woods or coppices, simply for the essential timber for use in the coachbuilding business. Timber was also bought from a variety of agents in towns throughout the same area as recorded in the Timber Register. For example, on 20th October 1902, Messrs Ellis & Everard of Hitchin supplied '*6 oak But*[t] *185 feet and 3 pieces of Ash*' at a total cost of £30:10s:0, including carriage, and on 1st September 1905 Messrs J Willmott & Sons of Hitchin supplied a quantity of Dry Oak at a cost of £26:12s:6d. Some timber suppliers came from much further afield such as Sadd & Sons of Maldon, Essex; Matthews of Ilford; Astell of Bedford; Francis Jacob of Huntingdon; Bell & Sons of Saffron Walden; and even Palmer of Dover. The types of wood purchased included elm, oak, deal, ash, beech, walnut and willow.

Carriage Range

Whether they were all original, or some copies of carriages built by competitors, readers may be surprised by the vast number of carriage designs offered to customers by Sanders during the early years of the twentieth century, designs created or copied by Frank himself, assisted by Lance. In the Sanders' archive there is a catalogue containing a diverse range of types. The mainstay of the business seems to have been the Governess Car, although even this type of carriage had many different styles, including the *'Eclipse'* and *'Special Eclipse'*, the *'Empress'*, the *'Encore'*, the *'Haileybury'*, the *'Hertfordshire'*, the *'Kingsbury'*, the *'Miniature'* and the *'Royston Insidecar'*. They were light and well balanced and sold in great numbers over a wide area. In addition there were Gigs, including the *'Liverpool'*; Dogcarts, including the Lady's *'Princess'* and *'Panel'* and the Ladies and Gentlemen's *'Stanmore'*, plus a four wheeled *'Stanmore'* Dogcart; Buggies, including the *'Westminster'*; and the *'Luggage or Station Car'* *'indispensable for the country house!'* There were also larger four wheeled carriages

Rubber tyres of all the best makes are fitted on the premises; no expense is thus incurred in sending wheels to tyresmiths as is usual with many firms who must obviously charge higher prices for this work.

Urgent orders for carriages have special attention, but no work is sent out until in a perfectly fit condition for use.

(3493)

Sanders' "Princess" Phaeton.

This carriage is deservedly very popular; it is stylish and its appearance suggests comfort that is hardly attainable in any other carriage. It is one that is eminently in favour as a lady's carriage, being easy of access and quite low. The concave panels project well over the wheels and the possibility of mud thrown from the wheels entering the inside is entirely overcome. It can be turned round in very narrow roads and is therefore specially suited for use in country districts. Its price with woodwork in natural colours, iron and steel parts black, is:—

Pony-size 38 guineas.   Cob-size 40 guineas.   Horse-size 43 guineas.

Unsolicited opinion.

We have well tried the Carriage and are extremely pleased with it. I am thoroughly satisfied with the way you have conducted the business and shall be very pleased to recommend any of my friends to buy from you.   MARDEN, KENT.

Finished specimens of this Victoria nearly always on hand.

Full specifications sent free of cost to any enquirer.

(6084)

Sanders' Fashionable "Queen" Victoria.

An elegant and luxurious carriage. Built and finished throughout in irreproachable manner and although quite moderate in price it will bear comparison with those sold by the best London houses. It is light in weight but at the same time perfectly good and durable and the builders confidently recommend it the best value obtainable. Built in two sizes.

No. 1 from 115 guineas.   No. 2 from 120 guineas.

Unsolicited opinion.   BAKEWELL, DERBYSHIRE.

Mr. ——— writes to say the Victoria arrived safely, although it did not escape the rain, but it has not suffered any ill effect from the wet.

Mr. ——— is entirely satisfied with its appearance, and its running seems very easy.

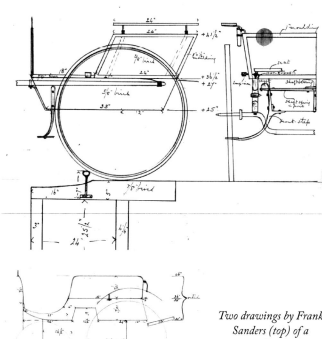

Two drawings by Frank Sanders (top) of a miniature buggy c.1905 and (left) of a Stanmore Phaeton, c. 1905.

like the Phaetons, including the *'Stanmore'*, the *'Princess'* and the *'Miniature Victoria'*; the *'Improved Runeasy Phaeton Dogcart'*; the *'Hertfordshire Bus Wagonette'* with a removable top in three sizes holding from four to eight passengers and the cheaper *'Bus'*; Broughams, including the *'Improved Circular-Fronted'* and *'Square-Fronted'*; Landaus, including the *'Miniature Canoe'*; and the *'Hertfordshire Victoria'* and the *'Queen Victoria'*. They were all made to the customers' specific requirements. Prices ranged from below 20 guineas for the basic Governess Cars, around 30 guineas for a dogcart, more than 40 guineas for a phaeton with approaching 140 guineas for a brougham or the largest bus wagonette and even more for a landau, plus the cost of suitable lamps and other accessories. (A guinea is one pound and one shilling or £1.05p.) The catalogue also offers *'harnesses made to measure in any size, hand sewn throughout, at short notice'*. Some examples from the catalogue can be seen elsewhere in this book.

There is a company brochure from 1908 in the Sanders' archive which promotes the diverse range of products and, on the front cover, states specialities include "*Landaus, Governess Cars and Motor Bodies*". It has a photograph of the first floor showroom where a large number of completed carriages are parked awaiting customers' purchase or collection. It contains warnings about *'Cheapness'* saying *'where cost is the primary factor with a coachbuilder, he obviously will use materials of a secondary* nature', 'Quality' saying 'no better ... proof of the quality of a carriage can be desired than the candid opinion of a disinterested user' and the 'Satisfaction' of their customers saying *A well-appointed carriage has many little refinements which cannot fail to favourably impress its owner and which reflect the attention bestowed by the builder who has studied his client's requirements*'. It also claims that Sanders 'have volumes of Unsolicited Testimonials ... from delighted owners'. The eighty such entries in the brochure include, referring to 'an Olympia Buggy', a quote reading *The little gig has arrived quite safely and is just what I wanted; it seems to run beautifully and I am very pleased with it. L.E.F. Studley, Warwickshire*' and, referring to a 'Special Eclipse Governess Car', the quote reads *We are both extremely pleased with the car. E.J.M. Shefford, Bedfordshire.*'

Day Books

There are a variety of Day Books in the Sanders' archive but only one relating to the sale of new and second hand carriages and motor bodies. This Day Book covers the period from 1906 to 1909 and includes details of over 500 bodies built for both horse drawn and motor chassis, the proportion of motor bodies gradually increasing year by year. I believe that motor bodies started to be built a few years before 1906, certainly the cab bodies were built in 1905 (see London Cab later). With no documentary evidence, I do not know the exact date but would

suggest that it was shortly after 1902 following completion of the new workshops. I have chosen some typical carriage and motor body commissions recorded in the Day Book, and elsewhere, and give details later in the book. Sanders' customers were by no means local with addresses in most counties of England and Wales, as well as many in London and as far away as Cape Colony in South Africa. Below I have listed the names of some of the customers from the index in the Day Book which includes well known Hitchin families and businesses, many members of

A Sanders' two-wheeled buggy, c.1910.

(Top left) A Sanders' Hansom cab, c.1905. (Top right) A Sanders' four-wheeled buggy (Bottom left) A Sanders' built hearse standing outside the Walsworth Road premises, c.1914.(Bottom right) A Sanders' buggy fitted with brass lamps, c.1910

the clergy, the military and the aristocracy (with some names sounding like characters from Trollope's *Barchester Chronicles*) and even other carriage builders, including the following:

The Honourable F W Anson
The Reverend Monsignor Bickenstaffe-Drew
Fisher Bowman Esquire
Colonel W St John Burke
The Right Honourable, The Earl Cardigan
Lady Castlereagh
Miss Cory-Wright
Messrs. S H Fordham
Heatly Gresham Engineering Ltd
Herts and Essex Public House Trust
Ernest Hankin Esquire
Lady Ilchester
Messrs J Jackson & Son
Mr E Logsdon
The Earl of Lytton
The Right Honourable, the Lord Viscount Melville
The Reverend E J Morris
Messrs Nicholls & Son
Mr Francis Newton
J Lovelace Ribbons Esquire
The Reverend J C Sparrow

Major H Bland Strange and coachbuilders
Messrs W C Windover, Turrill & Sons and later
Messrs Charles S Windover & Co.

Commissions – horse drawn vehicles

The vast range of carriages offered to customers has been mentioned. Six examples of a variety of horse drawn vehicles built by Sanders commissioned by its customers – both for personal transport or for commercial purposes – appear later in the book in the section entitled *"Commissions – horse drawn vehicles."*

Cape Colony

I mentioned earlier a customer in Cape Colony in South Africa. A copy of a letter in the Sanders' archive, sent by Frank to Alfred Keeler, possibly an agent or carriage dealer in the United Kingdom, enclosed carriage catalogues together with trade and private customer price lists. The letter clarifies prices by stating *'The prices given are what we expect from customers in this country, of course you will have to adjust them to the general conditions of things in S. Africa, the prices there are no doubt quite 35 p'c higher than here, for English carriages'.* It continues *'We have selected two excellent traps & shall pack them as one, all the parts will be easily put together & will be marked to avoid any mistakes.*

London Zoo

There was a family story that Sanders supplied a small number of their Governess Cars to London Zoo. This story was borne out by a copy of a Day Book entry in 1911. But did the zoo still have them? I decided to visit the zoo in Regents Park, London and try to find out and was accompanied by Priscilla Douglas. We were pleased to discover that the Zoo had indeed first received the carriages before the First World War and were told that they had been used to carry visitors around the zoo. The motive power had varied with ponies, llamas and even an ostrich being placed between the shafts. The ostrich had been trained in the private zoo of Anthony H Wingfield Esq. (later Sir Anthony) of Ampthill House, Bedfordshire where he kept a large number of animals which he trained to be ridden. In later years he was a member of the Zoological Society's Council and a generous benefactor of Whipsnade Zoo, including the donation of his exotic animals. The carriages were refurbished in the 1950s, the original varnished finish being painted over

(Above) A Sanders' Governess Car at London Zoo, an ostrich between the shafts and Keeper George Blore in attendance with his assistant Dexter c. 1913. (ZSL London Zoo collection)(Below)Sanders' Governess Car at London Zoo, with a llama between the shafts c.1914. (ZSL London Zoo collection.)

in bright colours. We were allowed to trawl through their photographic archive to search for photographs of the carriages and, whilst searching through 10,000 photographs (yes, ten thousand), we were successful in finding several photographs of Sanders built carriages drawn by both llamas and an ostrich, with Keeper George Blore in attendance, together with his assistant, Dexter. We were told that, as recently as 2008, the carriages were transferred to Whipsnade Zoo, in Bedfordshire, less than twenty miles from where they were made, and were subsequently sold. Sadly, it has been impossible to discover their fate.

Second hand carriages

In addition to new carriages specific Day Books record the sale of second hand, or 'nearly new', carriages such as *'a second hand square shape landau by Shanks painted green to Mrs Tyrrell of Ashwell for £55.0s.0d.'* or *'a small black four-wheeled Ralli Phaeton sold to Mrs Hubard for £22.1s.0d'.*

Early Motor Bodies

Readers may have limited knowledge of early motor cars. I hope the more knowledgeable will excuse a brief explanation. Many motoring 'firsts' are disputed but it is generally accepted that the first modern automobile was built by the German, Carl Benz, in 1885. By around 1895 the first motor cars were running on the roads of Britain, mostly imported but one-off vehicles started to be made around this time. The first series production motor cars were probably those produced by the British Daimler Company in Coventry, in 1897. The early vehicles had very rudimentary bodies, seldom more than a cover over the engine and somewhere to sit – often on top of the engine. By the turn of the 20th century, when individual manufacturers were beginning to build a series of similar cars on a standard chassis, fitted with an engine made either by themselves or bought in, the opportunity arrived for each customer to have a body made exactly to their own specification and to meet their particular needs. The obvious trades people to build such bodies were the many firms of coachbuilders who had been skilfully making carriage bodies for decades, in some cases centuries; most towns the size of Hitchin had at least one such company, possibly more. It is not surprising, therefore, that some of the early bodies for motor vehicles resembled those used on horse drawn carriages leading to the epithet 'horseless carriage'. Even with the start of mass production after the First World War, by such as William Morris (later Lord Nuffield), who studied the methods used by Henry Ford in Detroit, many manufacturers still made available complete running chassis for the customer's bespoke body to be fitted by their chosen coachbuilder.

A Belgian Métallurgique fitted with a Sanders' sports tourer body c.1914.

A Sanders 'Cambridge' side-entrance body fitted to a Cottin et Desgouttes chassis c.1908.

A luxuriously upholstered and appointed Sanders' landaulette c.1912

Commissions – motor bodies

Sanders offered a wide choice of passenger bodies to customers, including small two-seater open cars through to limousines and landaulettes (the spelling used by Sanders throughout their sales literature). The Sanders' archive has a catalogue for the years 1910-11 which includes various styles of coachwork like the *'Cambridge'* (costing 70 guineas up to 15hp and 75 guineas above 15hp), *'Cavendish'* and *'Newbury'* side-entrance touring cars, the *'Brighton'* (costing 172 guineas), *'Mentone'* and *'Surrey'* landaulettes, the *'County'* limousine and the *'Anglesey'* cabriolet (costing £233). The choice of chassis manufacturer upon which Sanders fitted bodies was also varied. British manufacturers included Morris, Bean, Napier, Daimler, Rolls-Royce and the Ford Model T made in Manchester. European manufacturers included Lancia, FIAT, Panhard, Mercedes (Mercedes-Benz was not created until 1926) and many Cottin et Desgouttes. American manufacturers included Hudson and Hupmobile. A photograph of the latter make, fitted with a Sanders' coupé body, was illustrated in the 20th December 1919 issue of *The Autocar* magazine and was brought to my attention by the late Tony Beadle. It should be borne in mind that these luxury vehicles often took four, or more, months to complete, not surprising when one takes into account that it was necessary to apply up to twelve coats of paint and varnish to give a good finish; the final coats of varnish being applied under the strictest workshop conditions to avoid blemishes. A helpful leaflet was given to owners warning them to wash mud from the bodywork, and finish off with a chamois leather, to help preserve the paintwork, a task often performed by the chauffeur.

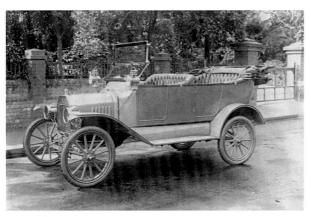

(Top left)A Sanders' Tourer body fitted to a Ford Model T chassis c.1910.(Top right) A Ford Model T chassis fitted with a different style Sanders' Tourer body, with the hood up, c.1910. (Bottom left) A Sanders' limousine body fitted to a Morris Oxford chassis c.1920. (Bottom right) A Sanders' landaulette body fitted to a Ford Model T chassis, c.1914, often bought by companies to offer for hire.

(Above left) The covered carriage truck used by Sanders to deliver vehicles to customers around the country using the railway network, one of two built in 1905 and photographed at Doncaster in 1913. (National Railway Museum/Science & Society Picture Library) (Above right) The Sanders 'works' Ford Model T van used before the First World War. (Below left) The Sanders 'works' van on an unknown chassis used in the 1920s. (Below right) The Sanders 'works' Series 1 Land Rover used in the 1950s.

(Above left) A Sanders' landaulette body fitted to a Cottin et Desgouttes chassis, c.1912. (Above right) A Sanders' saloon body fitted to a Rational chassis, c.1910
(Bottom left) A Sanders' Doctor's Coupe body fitted to a Bean chassis, c.1920. (Bottom right) A Sanders' limousine body fitted to a Napier chassis, c.1914

(Above left) A Sanders' landaulette body fitted to an unknown chassis c.1912, standing outside the Walsworth Road garage.
(Above right) A Sanders' landaulette body fitted to an unknown chassis.
(Left) A Sanders' landaulette body fitted to an unknown chassis, c.1910.

44

Five examples of a variety of motor bodies built by Sanders, commissioned by its customers for their personal transport or, in the case of the Adams Manufacturing Co., to be supplied to their customers, appear later in the book in the section entitled *"Commissions – motor bodies."*

Cooper Cars

The main source of information about the building of Sanders' motor bodies is their sole surviving Day Book for 1906 to 1909. Not only is there little documentation about vehicles built earlier than 1906 but, disappointingly, the same deficit applies to the period subsequent to 1909 when most of the motor bodies were actually built. You will see a photograph earlier in the book of a Rolls-Royce bodied by Sanders but there are no details in the Sanders' archive about who commissioned the body. However, there are many photographs of a variety of vehicles built by Sanders, although the make of chassis is not always recorded. With the help of experts at the Vintage Sports-Car Club many have now been identified; these include the Cooper car. Further information was found in Nick Georgano's encyclopaedia.

The Cooper cars were built by traction engine manufacturers, the Cooper Steam Digger Co Ltd. of King's Lynn, Norfolk, founded by Thomas Cooper in 1894. At one point he held more patents than any other single person in the country. The cars

Three of only six cars built by the Cooper Steam Digger Co Ltd of King's Lynn, Norfolk were fitted with bodies made by Sanders. Note the distinctive narrow rim of the radiator.
Above: A Sanders' landaulette body with the door open to show the sumptuous upholstery and occasional seat, c.1911.
Page 46: A Sanders' landaulette body with no protection for the driver, c. 1909.
Page 47: A Sanders' limousine body with ornate rooftop luggage rack, c.1911.

COACH BUILDERS
RALPH E. SANDERS & SONS
HITCHIN & ROYSTON

were made between 1909 and 1911 and were destined for very select customers, exclusively directors of the company and their friends. In fact only six cars were built and I know for sure that at least three were bodied by Sanders; photographs of all three survive. One car was exhibited at the 8th International Motor Exhibition at Olympia in 1909. The cars were powered by an ingenious 22hp 4-cylinder 3,233cc piston-valve 2-stroke petrol engine, designed and made by the company, and fitted with a 3-speed gearbox and 2-speed rear axle, giving six forward speeds and two reverse. One of the six cars, a tourer registered BM 1153, survives today and is stored at Cooper's successor company, the Cooper Roller Bearings Company Ltd., who still have a factory in King's Lynn, now part of a multi-national operation. I went to see it hoping it would become the only surviving Sanders' bodied motor car but, when I opened the rear passenger door, I was disappointed to find that the chassis plate read *"Maythorn & Son of Biggleswade"*, Sanders' local rivals just across the boundary into Bedfordshire. I have also seen on the internet a photograph of a Cooper 2-seater drop-head coupé registered EB 2509 but I do not know who made the body, although the wide scuttle is a distinctive feature of a Sanders' motor body.

London Cab

In 1905, Sanders made the first bodies for the London taxi

Two Rational cabs built by Sanders outside the Savoy Hotel, The Strand, London in 1907. (© TfL from the London Transport Museum collection).

cab – a controversial statement? Well, the experts will certainly take issue with the description 'taxi cab' as the word 'taxi' did not come into common use until 1907 at the earliest, the word taxi being an abbreviation of the taximeter compulsorily fitted to each cab from that date and checked by the Public Carriage Office. However, although an electric powered cab had been introduced onto the streets of London as early as 1897 – the unsuccessful 'Hummingbirds' introduced by Walter Bersey, General Manager of the London Electrical Cab Co – I believe that amongst the first petrol driven cabs were those made by The Bassingbourn Iron Works, later known as Heatly-Gresham Engineering Co Ltd, of Bassingbourn, Cambridgeshire, makers of the Rational motor car, who later moved to Letchworth. I confidently make this claim for Sanders because I have a copy of a letter written by Frank and sent to J.N. Queenborough, Assistant Editor of *The Garage & Motor Agent*, dated 13th August 1942, in which he says *'In 1905 we built the bodies for 6 of the first London Motor Cabs. This was mentioned in an old contemporary, the Coachbuilders & Motor Car Journal, in that year, and we have the Day Book entries available; the first Cab caught fire when passing over one of the London bridges, and we also see there is an entry in the Book of its repair by our firm.'* Frank offers to send them the Day Book in a later letter. Whether he did or not is unknown but it does not survive in the Sanders' archive. Could it have been sent to the Assistant Editor as offered and got lost or perished during the London bombing? However, the Day Book for 1906-09 does record the later supply of a further two new motor cab bodies to 'Heatly Gresham Engineering Company of Letchworth', the first on the 28th March and the second on the 28th April. The entry reads *'built and finished as per specifications supplied, mounted upon chassis complete for the inclusive sum of £55.0s.0d each'*. The full title of the magazine referred to by Frank was *The Coach Builders', Wheelwrights' & Motor Car Manufacturers'*

Art Journal and, at the foot of page 165 of the June 1905 issue, it says *'Messrs. Ralph E Sanders & Sons, coach builders, of Hitchin, have completed the first Motor Cab for the new London cab company'*, lending further support to my claim. London taxi cab expert, Bill Munro, told me *'there was a report in Motor Traction magazine (I think) whose reporter took a Rational cab from Northumberland Avenue, off Trafalgar Square, to Brighton. The journey took 3½ hours; Purley was reached in 90 minutes, which would be impossible in today's daytime traffic!'* In the booklet *Seventeen Taxis*, written by David Hamdorff and published in 1983, there is a photograph of a Rational cab standing outside the gates of Wimpole Hall, close to the village of Bassingbourn, but, although it also includes a picture of the Rational saloon car attributed to Sanders, it sadly makes no mention of Sanders being the cab's coachbuilder. I also have a photograph of two Rational cabs on the London streets kindly provided by the London Transport Museum.

Commercial vehicles

Other photographs in the book show some of the commercial and public service vehicles built by Sanders. These not only include the chassis fitted with van bodies for the company's own use but several others, including the vans made for the fondly remembered Hitchin confectionery company, Garratt & Cannon Ltd., who had premises in Bancroft, Hitchin. Larger vehicles were also built, like

(Above left) A Sanders' van body fitted to an Albion chassis and supplied to Hitchin confectioner Garratt & Cannon Ltd., c.1920. (Above right) A Sanders' van body fitted to a Bean chassis and supplied to confectioners Cannon's for their Cambridge works, c.1920. (Below left) A Sanders' pick-up truck body fitted to a Model T Ford chassis, c.1914. (Below right) A Sanders' Wagonnette body fitted to an Albion 15hp chassis, c.1915

(Above left) A Sanders' charabanc body on a Model T Ford chassis, c.1920. (Above right) A Dennis single-deck bus built by Sanders for Road Motors Ltd and used on the route from Letchworth, via Hitchin, to Luton, c.1920. (Below left) A Sanders' charabanc body on a Lacre chassis made in Letchworth, Hertfordshire, c.1920. (Below right) A bus or coach body built by Sanders and fitted to a Talbot chassis, c.1920.

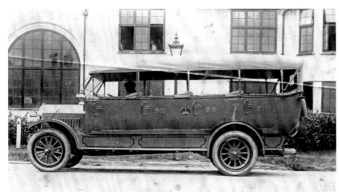

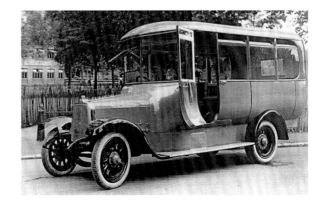

(Above left) A Sanders' tourer body fitted to an unknown chassis, c.1910, standing outside the original Walsworth Road workshop. (Above right) A Sanders' landaulette body fitted to an unknown chassis, c.1912, standing outside the brick built car showroom. (Bottom left) A Sanders' limousine body fitted to a Daimler chassis, c.1914. (Bottom right) A Sanders' cabriolet body fitted to an Italian FIAT 501 chassis, c.1920

the drop-sided truck body fitted to a Ford Model T chassis but even larger still were the buses and charabancs, like the single-deck bus built for Road Motors Limited who operated the route from Luton to Letchworth via Hitchin.

Second hand car sales

I know from photographs that second hand cars were also sold by Sanders during this period; some of them can be seen on the front cover lined up outside the Walsworth Road

premises. Thanks to the late Tony Beadle I also have a copy of an advertisement in *The Motor* magazine dated 21st September 1915 which shows a photograph of a Sanders flush-sided cabriolet body fitted to a late 1913 Enfield 12-14hp chassis. The description includes '*…carry three inside, occasional seat for extra passenger, Stepney* [spare] *wheel, Rotax electric lighting set, luggage grid, speaking tube, painted purple lake, upholstered Bedford cloth. Run less than 2,000 miles'*. The asking price was £325.

Car repairs and minor transactions

In the Sanders' archive there is also a specific Day Book, dated from 1906 to 1909, relating to the repair of carriages and motor bodies. The repairs are often minor, such as repairing the axle on Miss Hankin's cart at a cost of £1.2s.0d, but sometimes are full rebuilds, such as the work carried out on a landau for Mrs Sowerby of Kings Walden at a cost of £21.14s.0d. In a trade magazine, '*Garage and Motor Agent*', a small eight page paper issued monthly and circulated gratis, dated 9th October 1943, there is an entry under the theme '*Motor Trade Family Album*' about the Sanders business entitled '*Only Two Principals in 68 Years*', being Ralph and his son Frank. As well as a brief summary

A Sanders' cabriolet body fitted to a 1913 Enfield 12-14hp chassis, shown either when new or when offered for sale second hand by Sanders in 1915 for £325.

of the history of the business, it refers to *"the erection of a motor repair garage in 1906-7 … one of the earliest brick built structures in the country specifically designed for motor chassis repairs.'* It also records that *'Well ahead of the general change-over from canned petrol to bulk storage, the firm installed a 2,000-gallon Bowser tank and pump during the 1914-18 war.'* It is believed that the tanks may still be under the pavement.

In addition, it records that *'the first booked entry of a repair job goes back to May, 1903, when a gearbox of a Belgian car was overhauled for Lord Alwyne Compton.'* This last statement encouraged me to find out more about this aristocratic motorist. He was the third but second eldest surviving son of Admiral Compton, the 5th Marquess of Northampton, and one of his uncles was Bishop of Ely. He was educated at Eton and had a distinguished military career, was aide-de-camp to Lord Ripon, Viceroy of India, took part in the campaigns in the Sudan and was awarded a DSO in the Second Boer War. He was MP for Biggleswade from 1895 to 1906, a town only about 10 miles from Hitchin. He died in 1911, aged 56.

There is further evidence about this incident in the autumn 1960 issue of *Hertfordshire Countryside* which contains a letter from a Mr E H Davies of Alberta, Canada, then aged eighty. He claimed to be *'the only mechanic employed on the first repair job on a car ever undertaken by Messrs R E Sanders and Sons in 1903, and then only as a helper, since I knew practically nothing of the mechanism of a motor-car before that time.'* He continues *'Lord Alwyne Compton was on his way to camp with the Northants Yeomanry at Hinchingbrooke Park, Huntingdon, when by some means the gear shaft became slightly twisted so that it was impossible to change the gears.'* He says that the chauffeur must have seen the Sanders Garage sign in Bridge Street, pulled up and tooted for attention. Lord Compton continued his journey by train but his chauffeur, a Frenchman, said that he would carry out the repair with the aid of some tools and a helper, Frank asked him to help. Together the two men removed the gears from the shaft with a block of hardwood and a blacksmith's hammer and the chauffeur took the shaft to London by train for repair. He added *'Had I had the mechanical knowledge I have now, I could have taken the twist out right there.'* The following night he met the midnight train and, after a *'midnight lunch'* with Frank at the Boot public house, next door to the garage, he replaced the gears on the shaft and re-fitted it to the car. He finished the job at 4.30 a.m. He concluded *'The chauffeur invited us to go for a ride, my first in a motor-car. He took us out on the Bedford Road, and when he opened up to forty-five miles per hour he had us both hanging on to the sides, particularly when we went over a culvert.'* It was an exhilarating experience for the young man.

In addition there is a further Day Book, dated from 1902 to

1904, that records minor transactions, such as: *'one tin of burning oil and a wick for Miss Dawson of the Market Place costing 7d; a pair of trouser clips for Mr Hankin costing 2d; the hire of a bicycle by Miss Kirkland costing 6d; and adjusting Miss Dollimore's pedals costing 2d.'* Where the name of a customer is not known a very brief description is given, such as *'working man'*, *'cyclist'*, *'coachman'* or, most commonly, *'stranger'*, sometimes followed by *'young girl'* or *'from Stevenage'*. Another record book includes some very detailed drawings by hand of various types of carriages giving precise measurements.

The Day Books that do survive give further fascinating information about the building and repair of a wide range of passenger and commercial vehicles but lack of space does not allow me to record any more.

The First World War ended a golden age for the wealthy Edwardians and the carriage and car buying trade would never be quite the same again. Consequently there was less demand for coachbuilt carriages and cars, Sanders' core business so far – diversification was to be the order of the next decade.

The "Cambridge" Side-Entrance Car.

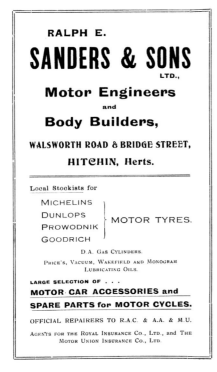

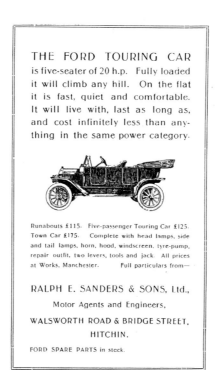

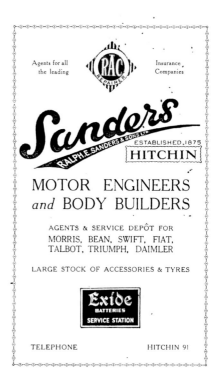

Two Sanders' advertisements from 1915 with the one in the middle being. for 'The Ford Touring Car', based on the Ford Model T.
The third advert (right) is from 1921

The Roaring Twenties:
the business that offers everything

There is a dearth of material for the period between the wars. There are no Day Books or any other documentary records belonging to Sanders, very few photographs and limited, mostly second or third hand, personal recollections or anecdotes. However, it seems clear that the initial post war boom ensured that Sanders had the financial basis to continue to build its fortunes enabling it to add new buildings and provide further services. The continuing advances in the technology of transport did have an impact upon the services offered, particularly an adverse effect upon the manufacture and sale of horse drawn vehicles. A positive effect was that the widening range of those services must have helped the business weather the downturn that occurred once that initial boom had lost steam.

Walsworth Road

The last major building to be constructed in Walsworth Road was a car showroom. You will recall that, in the first decade of the century, two-storey brick buildings were built at each end of the site and that, adjacent to the right hand building, the one closest to the station, there was a building, opensided on the left, under which cars were displayed for sale. This structure was retained and became enclosed by a further building on the site, to its left and filling in the gap, being a single-storey car showroom with a basement. It was constructed with a pitched tiled roof, large

The Walsworth Road premises, now including the new car showroom built in the 1920s

58

glass skylights and a façade in the Art Deco style using stone and brick, central Crittall type steel and glazed double doors for customer access and sliding metal doors each side to enable cars to enter and leave the showroom. (The remaining Crittall doors and windows were replaced with modern steel and UPVC in 2013). The work was carried out by John Willmott & Sons of Hitchin at a cost of £1,629:l0s:2d. In front of the showroom was a range of petrol pumps supplying fuel to customers. Inside there was an office with a glazed window allowing sales staff to view activity in the showroom. The remaining buildings provided most of the facilities required for body building, all catered for in-house. There were carpentry workshops, an area for painting, varnishing and upholstering the bodies and for their assembly; in the case of motor bodies, this included the equipment to enable them to be attached to the motor chassis. There was also a forge making most of the metal parts required in the construction of horse drawn vehicles, including the chassis, axles, wheels and springs. The forge also made and repaired springs for motor vehicles and other fixings. In addition there were workshops for the repair of motor cars and an area for the sale of motor accessories and spare parts, including tyres; batteries for motor cars were also made. Later in the twenties Sanders advertised themselves *as agents and service depot* for such manufacturers as Morris, Bean, Swift, FIAT, Talbot,

Daimler and Triumph, the latter becoming significant in the company's history, as mentioned later. ('Agents' usually obtained their cars from the local 'dealer' for a particular manufacturer and the dealer obtained their stock from the area 'distributor' who obtained their vehicles direct from the manufacturer). In fact, the business offered everything a customer might require whether his transport was horse drawn or motorised - or indeed pedalled, as they also built, sold and maintained bicycles.

Bridge Street

I know that, probably in 1915 but at least by 1920, Sanders were again in occupation at number 13 Bridge Street and that second-hand cars were displayed and sold from this new purpose-built building. An evocative photograph, taken in this period, shows cars standing outside; two of them have Hertfordshire registration numbers, one including 'AR' and another 'NK'. Boards outside the garage advertised *'Gargoyle Vacuum Motoroils'* and *'Prowodnik "Columb" Tyres'*. Through the windows could be seen motoring accessories and lubricants displayed on shelves. Later, advertisements for *'Lodge Plugs'*, *'Osram Lamps'* and *'Ferodo Brake Linings'* appear in the upper windows. A further sign declares *'Cars for Hire'*. There is also a petrol pump, close to the kerb to the left of the garage, to supply fuel for customers' cars. Tenants continued to occupy the other Bridge Street properties, including, from 1923,

number 14, the former Royal Oak public house (previously known as The Boot) which had been purchased that year.

Shareholders

Frank's shareholding increased during this period, as did that of Lance, Doris, Bill and Roy. Bernard received his first allocation of shares on 3rd December 1923 and further shares were allocated in 1924 and 1925. I assume that Ralph was relinquishing some of his shares. The company was still very much a family business.

Personnel

I know that Frank continued to be in charge of the business but, in 1921, he was joined by his son, Bernard, straight from school at the age of 15. Frank's brother Lance was also involved in the business. His expertise as a carpenter and joiner was utilised in the workshops where the carriage and motor bodies were built. He also dealt with the upholstery of the vehicles. Frank's wife, Alice, was the company's first bookkeeper, replaced later by Frank's sister, Doris, possibly when Alice's health began to fail. Alice died in 1931.

(Left) Sanders' purpose built car showroom, at No. 13 Bridge Street, with a car inside for sale and customers' cars receiving attention outside c.1920s – note the petrol pump.

The names recorded on indentures for apprenticeships for this period include Alec Stanley Day and Albert John Targett from 1925 and William John Taylor and Thomas Russell Barker from 1927. Again, how long they remained with the company and what skills they learned is unknown.

Central Wireless Stores

It is apparent from reading the opening chapter that the Sanders family were always willing to find employment for their family members, even distant ones. Ralph's two soldier sons, Bill and Roy, survived the First World War and, when discharged, both joined the family business. It is not known what Bill was employed to do but his period of employment was, in any event, very short. By the early 1920s he had become a farmer as Ralph had arranged for him to farm Lordship Farm in Cottered. It was also around this time that Ralph diversified his business by opening a shop specialising in domestic electrical equipment to exploit the use of electricity now being supplied to many homes in the town. Whether he did this to create a job for Roy, or planned to do it anyway, is not known. In 1919, the business opened at number 7 High Street, Hitchin, a small building between Perks & Llewellyn's, the pharmacists, and the Cock Hotel. Around 1930, Sanders vacated these premises and moved the business, now called the *'Central Wireless Stores'*, to Bancroft

to occupy part of a building that had started life as a medieval hall house. The High Street premises were then occupied by the Cock Hotel when their original larger building next door was demolished to make way for the new Woolworths store (now Boots the Chemist). To explain more about this business, its trade and its staff, Roy's daughter, Joanna Cooper remembers:

"My father Royston (Roy) Sanders had always been fascinated by electricity. This latest venture, though still part of the business, was, I always understood, my father's own enterprise. The shop had a manager, Alfred Brown, with Miss Bishop as his assistant. Mr Brown was also the buyer from 'reps' who came in during the week and who readily bought the latest gadget of the day. My earliest recollection of the shop was in 1937 when I was six years old. My father always walked into the town on a Saturday afternoon at about 4:30pm. He would discuss the business of the week with Mr Brown and collect the day's takings to put in the bank. Occasionally he would ask if I would like to go with him. This was an expedition I always enjoyed, not least because of a visit to the sweet shop in Hermitage Road on our return!

————————

(Left) Sanders' original premises for their electrical business, at 7 High Street, set up after the First World War by Roy Sanders, with (from left to right) Roy, manager Alfred Brown and Lance. The building survives as the Cock Hotel. (Right) Sanders' electrical business after the move to Bancroft, and now named the 'Central Wireless Stores', with the manager, Alfred Brown, and his assistant, Miss Bishop, standing in the doorway c.1930s

One particular occasion I remember clearly. The shop had a heavy oak framed glass door with a big brass latch and a bell that would sound to alert our entrance. The interior of the shop, with its shelves full of wirelesses of all shapes and sizes, lamps and clocks, irons and torches and vacuum cleaners and much more, was like Aladdin's Cave to me. Mr Brown, with his beaming smile, was behind the counter radiant with pride at his latest prize exhibit on the end of the glass counter. That position was always the place for the latest device. I was aware that my father had stopped stock still in his tracks at what he saw. I was in front of him. The subject of our gaze was a table lamp clock (for lack of a better description). The shade at the top looked as if it had been made of glass with coloured bobbles, which reflected on to the ceiling. Under the light bulb was a square or rectangular chrome framed clock with analogue numbers. It was the stand however that had stunned my father. It was in the form of a glass fish bowl with what appeared to be a complete shoal of look-alike goldfish swimming round the circumference at great speed. This was not all, because the lighting inside changed from blue to green from top to bottom. I can't remember whether I looked up at my father - in hindsight I hoped I didn't because he might well have burst out laughing and I know he would not have wished to offend Mr Brown. I clearly remember the silence being broken hearing my father say 'you'll never sell that.' I left the two of them talking and went to investigate my other fascination, the shop itself. I always thought it would make a lovely house. I knew it must be very old because the wall that divided the front of the shop to the back room was made of old beams and at about my father's height was a balcony. There was easily enough space to walk within it like a small passage, but where did it go? What was more of a problem was how did you get up there in the first place? I went into the back room and looked amongst the packing cases to see if I could find a little staircase or steps. I cannot remember ever asking my father to help me out with an answer. It was only recently that I learned that, if I had ventured beyond the packing cases, I would have found the original wooden spiral staircase that led up to two rooms – one used for wireless repairs and the other for storage. When we left the shop my father usually studied the window display. On this particular day there was a life size cardboard cut-out of a family with the father holding a torch with the caption 'Let Exide see you home.' As we stood I asked 'Why did you tell Mr Brown he would never sell that lamp with the fish?' My father replied with a laugh and the usual twinkle in his eye, 'Because I knew he would.' My father always admired Mr Brown's selling prowess hence his remark. Many of the items were not of my father's taste but Mr Brown knew his customers best! At the start of the Second World War in 1939, my father joined the army for the second time in his life. My visits to the shop were either with my mother or older brother. There were not the exciting gadgets now – more utilitarian items geared to wartime, such as torches and lamps with shields on them so that the light could not be seen overhead by aircraft. Customers were usually women carrying accumulators in their shopping baskets paying sixpence to have them recharged. These were used as extra sources of energy for small radios. With the end of the war came the return of the television service and we at home may well have had one of the first post-war television sets. The wooden cabinet was about three feet tall with a slanted top revealing a three-inch screen. To us it was fantastic. With every new improvement Mr Brown had the latest set in stock. With the move to modernise the town centre the old Sanders' Central Wireless Stores was demolished in 1958; sadly, none of the old medieval buildings were retained."

Whilst researching her memories Joanna made contact with a former employee, Vic Martin, who had worked at the Central Wireless Stores performing all sorts of duties, including as an aerial fitter. He had married the boss's daughter, Betty Brown. Vic remembered buying a *'cat's whisker'* (a thin wire) for his crystal set, an early type of radio receiver used with headphones, from the shop in 1929 and still owns a Marconi wireless set bought from the shop. He recalled a time when he was serving in the shop, he thought on his own, and another local electrical dealer, Stanley Lee, came in wanting to buy some batteries. Vic resisted a request for a trade discount which impressed Mr Brown who was actually only in the next room. Vic was also able to confirm the names of the three men in the photograph (taken by local photographer Minnis) outside the High Street shop; they are Roy on the left, manager Alfred Brown in the middle and Lance on the right.

Bus Garage

For more than eighty years the first large building that greeted you as you entered Hitchin town centre from the south was the bus garage but did you know that there is a Sanders' connection? Frank's local knowledge of land and property in the town may have led him to a speculative venture to develop land in Bridge Street or he may have been approached by an individual or a company, local or from further afield, looking for new premises from which to carry out their business. On Jonathan Wilkins revealing website about the bus garage I read that, in 1925, Frank purchased the premises occupied by J Williams & Sons, Iron and Metal Merchants, behind his new car showroom at 13 Bridge Street but for what purpose is not known. Neil Thomson, whose father established a motor business in Queen Street in the 1930s, suggests that the site was purchased to enable a large new workshop to be built for G W Lawrence Ltd, coachbuilders of Codicote, who wanted to expand their business into Hitchin. This did not happen; perhaps they did not have the funds to meet payment for the project after all. They later purchased the existing coachbuilding business of the late John Cain in Queen

Street. The HUDC rate books record that, on 25th April 1927 and 1928, there was *'building land'* behind number 13 Bridge Street owned by Sanders and no rates were paid. In February 1928, a planning application was made for a motor garage to be built on the site. Subsequently, although the specific date is not known, a large garage and workshop was built. Whether it was always intended as a bus garage is unclear; there was certainly a need for one as the local bus companies had outgrown the use of the yard at nearby Prime's Garage in Queen Street.

In the next surviving rate book, for 30th September 1930, it is recorded that there was a 'garage and premises' (and a urinal) on the same site, behind number 13 Bridge Street, occupied by D & E Petty. This information is repeated on 30th September 1931 and 1932, although, by then, it had been allocated street number 22. They also occupied

A plan, probably made in the late 1930s, of the Hitchin Country Bus Garage built by Sanders in Bridge Street (including an air raid shelter). The small workshop, which extends to the left, butts up to the rear of Sanders Garage.

A view in 1939 of the bus garage built by Sanders at the foot of Hitchin Hill when let to the London Passenger Transport Board.

number 14 Bridge Street at this time, described as a 'warehouse'. But who were D & E Petty? I have seen an advertisement for this company which describes them as *'Passenger Vehicle Specialists'* and shows an AJS Pilot 24-seat coach as supplied to Express Motor Services of Grays, Essex. Further research has revealed that they built highly regarded bodies for buses and coaches, particularly for Wolverhampton chassis makers AJS and also some for the Gilford Motor Co of High Wycombe. *The Commercial Motor* magazine, of 11th November 1930, records *'From the foregoing it will be seen that Messrs D & E Petty are practical bodybuilders who*

thoroughly understand the requirements of coach and bus operators and do everything possible to meet them in an efficient, but inexpensive, manner'. They also had premises in Cheshunt, Hertfordshire. Under what terms they occupied the Bridge Street garage, or what rent they paid, is not known but it seems as though the difficult years of the Great Depression took their toll as the *London Gazette*, for 11th January 1935, records that, in 1933, Daniel and Edward Petty filed to dissolve their partnership. They would not have been competitors for Sanders as the coachbuilding side of their business had, by then, almost ceased.

By that year D & E Petty must have vacated the garage because, on 26th August 1933, Frank entered into a tenancy agreement with the London Passenger Transport Board (LPTB), created only four months earlier, on 13th April 1933, at a rent of £225 per year. The LPTB absorbed many privately run bus companies in the London area within a radius of approximately 30 miles. The tenancy continued until 19th October 1953 when the successor body, the British Transport Commission, purchased the site from Sanders for the sum of £11,400. The garage continued to be used to store Green Line and green painted London Transport buses, the entrance being heightened in 1954 to take double-decker buses, until it closed on 29th April 1959 when all buses were transferred to a new depot in Stevenage.

Knebworth Estate

The various sources of timber procured by Ralph were mentioned earlier. One of those sources was the nearby Knebworth Estate. Clare Fleck, Archivist at Knebworth House, brought to my attention correspondence between William Wilson, Estate Manager, and Ralph around 1919 and 1920. The business relationship seems to have been a delicate one but the letters are consistently formal and polite, although their tone is often forthright and stern. Some of the correspondence reads almost like veiled threats, including the one sent to Ralph about the injury to a horse, owned by a Mr Smith of Manor Farm, Knebworth, caused by a felled oak, purchased from the estate by Ralph for £13, not having been trimmed by his men, adding that, if the horse does not recover, Mr Smith will claim from Ralph the amount the horse had cost him £78.8s.0d. On another occasion Wilson wrote to Ralph to tell him that, if he didn't remove, *'this week'*, the trees that he had felled some months earlier, with *'the tops still on'*, a Mrs Reynard of Stevenage and a Mr Browning of Half Hyde had threatened to take action against him for damages. In a similar vein Wilson wrote to Ralph saying *'I have again to direct your attention to the very considerable damage your haulers are doing to our fences here. Within the last few days they took down a 10ft. length of the Park oak fence and left it down. The consequences were that the cattle got out and damaged Mr Cain's growing crops about which he is naturally much annoyed. I must charge you for reinstating the fence and our men's time."*

Further letters, certainly unveiled threats, accuse Ralph of removing more trees than he had purchased. One, written in very strong terms, says *'I give you formal notice on behalf of the Earl of Lytton that I claim damages against you for felling the trees which were not sold to you. The damages claimed are £50 and failing an amicable settlement the case will be taken to Court, for this kind of business must cease'.* Several subsequent letters followed but, when they petered out, I assume the matter was resolved. Wilson also complains in correspondence about the delay by Ralph with the supply of metal fence rails, presumably made in the forge at the Walsworth Road garage. The impression that I have formed of Ralph is that his reaction to these confrontations was likely to have been a shrug and that any dispute would be satisfactorily dealt with after a face to face discussion. To be fair to Ralph only the correspondence sent to him has been seen, his replies, referred to by Wilson, do not survive. On a more pleasant note one letter from Wilson to Ralph commences *'I was sorry to hear you are suffering from rheumatism and hope you will soon be better'.* Was Ralph a bit of a rogue in business matters or was he simply a hard-nosed business man?

There was one final and new expansion of the business at the end of the decade but I will tell you about that in the next chapter as it heralds a major change to the core operation of the business.

The Art Deco style car showroom built in the 1920s. Note the petrol pumps with long swing arms which extend over the pavement.

Challenging Times:
coping with the Great Depression and the Second World War

For this period, roughly starting from the time of the Wall Street Crash of 1929 up to just after the end of the Second World War, there is the least material in the Sanders' archive, although there is some first hand anecdotal history from current family members, former employees and those who had dealt with the company, including past customers, suppliers and local residents.

Shareholders

Following the death of Ralph in 1933, his shares in the company were divided amongst his 10 children and his grandson, Bernard.

Coachbuilders in the 1930s

In the wider world, the Wall Street Crash of 1929, followed by the Great Depression of the 1930s, severely affected the business as less money was available for spending on luxury coachbuilt carriages and cars. There was increasing competition from established motor manufacturers of complete luxury cars, already fitted with a body of their own manufacture or made by a particular coachbuilder with whom they had an arrangement or possibly even owned. These included Alvis, who relied on coachbuilders in the Midlands; Armstrong Siddeley, whose in-house coachbuilder was Burlington; Daimler mainly used Hooper & Co; Invicta had an arrangement with several coachbuilders in the London area; Lagonda used E D Abbott Ltd and others, with the later Frank Feeley designs built in-house; Lea-Francis favoured Avon and Cross & Ellis; Talbot made bodies in-house and Bentley, from 1931 owned by Rolls-Royce, were associated with H J Mulliner & Co, Park Ward and Vanden Plas amongst many others. In addition, higher volume British motor manufacturers were starting to build good quality cars at affordable prices such as Rover, SS (the Jaguar model name, introduced in 1935, becoming the company name post war), models from the Rootes Group like Humber, Hillman (independent until 1931) and Sunbeam (independent until 1935) and the larger models from Austin, Morris, Riley, Wolseley, Standard and Vauxhall. Fewer manufacturers offered chassis for the fitment of bespoke bodies, although there was, of course, Rolls-Royce, whose chassis were bodied by Sanders and who didn't produce their own in-house bodies until the introduction of the Silver Dawn in 1949.

The Business

As far as I know most of the building work on the original site in Walsworth Road had been completed by the 1930s. The company continued to be run by Frank until his death in 1949, assisted by Roy and with increasing help from his son Bernard. Bernard's own son, Robert, although a trained engineer, was

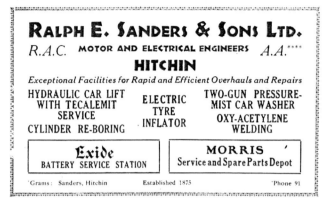

RALPH E. SANDERS & SONS LTD.

R.A.C. MOTOR AND ELECTRICAL ENGINEERS A.A.****

HITCHIN

Exceptional Facilities for Rapid and Efficient Overhauls and Repairs

HYDRAULIC CAR LIFT WITH TECALEMIT SERVICE CYLINDER RE-BORING

ELECTRIC TYRE INFLATOR

TWO-GUN PRESSURE-MIST CAR WASHER OXY-ACETYLENE WELDING

Exide
BATTERY SERVICE STATION

MORRIS
Service and Spare Parts Depot

'Grams: Sanders, Hitchin   Established 1875   'Phone 91

A Sanders' advertisement in a local paper in 1931 detailing the wide range of services offered to their customers - and potential customers.

(Top) The stores (Right) The tyre stores in Walsworth Road around 1930.

not employed in the family business. The demand for horse drawn carriages was, by now, infrequent, as indeed were orders for bodies on motor chassis. Sanders had never exclusively built bodies for the chassis of one particular manufacturer, or even from one country of origin, so they had no continuing link with a particular manufacturer. In any event, the repairing of the running gear of cars, rather than only dealing with their bodywork, had become the main occupation of the business combined with the sale of second hand cars and commercial vehicles of all types. Car ownership was increasing rapidly but competition to sell cars had also increased with the rise of small non-franchised dealers setting up business in the town to sell second hand cars. However, the range of services offered by Sanders was impressive. The *Hitchin Official Guide* for 1931 states '*Messrs. Sanders & Sons have the latest approved labour-saving plant installed at their Works such as Hydraulic Car Lift, Automatic Tyre Inflation, Overhead Travelling Crane, Cylinder Reboring Outfit, Mist-pressure Car Washer etc., and are thus able to offer car owners many advantages which are not often found in country Garages. It also maintains the Exide Battery Station and Tecalemit Services for the town*'. (Tecalemit developed a system for lubricating vehicles using high pressure lubrication). Sanders were also agents for motor insurance.

It is recorded that Sanders had been a retailer of a variety of makes of new cars, by arrangement with local dealerships, but they had not, so far, taken on a dealership themselves. There is a family story that the company had been offered the Austin franchise but they refused it, the reason now forgotten and later much regretted. However, Sanders did decide to embrace the growing demand from customers who wanted complete good value new cars by taking on the Triumph dealership selling the small Super Seven saloon which had been introduced in 1927. The exact date that the franchise commenced is not known but

is believed to have been in the late 1920s because, in 1929, Sanders were selling the Super Seven for £149.10s.0d. Sanders also took on the agency for Commer vans and trucks as shown on an advertising hoarding on the railway bridge at Little Wymondley (now by-passed) which passed over the main A602 road leading from Hitchin to Stevenage. On the other side of the bridge another hoarding declared 'SANDERS' GARAGE - HITCHIN 2½ MILES - THE SERVICE DEPOTS OF THE DISTRICT'.

By the mid thirties Triumph's general

An advertisement for the Triumph Super Seven sold by Sanders on their appointment as Triumph dealers c.1930

*Advertising on railway bridges. (Right) A hoarding advertising the whereabouts of Sanders' Garage on the railway bridge as you enter Hitchin from the north east, c.1930s. (Bottom left) Sanders' advertising hoarding, promoting Commer vans and trucks, on the north side of the railway bridge in the village of Little Wymondley, c.1930s. (Bottom right) Sanders' advertising hoarding, promoting their range of services, on the south side of the railway bridge in the village of Little Wymondley, c.1930s.
Back cover: Sanders' hoarding on the railway bridge as you exit Hitchin advertising the 'Triumph Gloria Vitesse' saloon sold by the company, c.1935.*

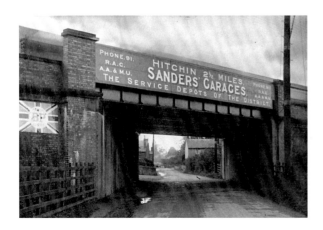

manager, Claude Holbrook, decided that the company could no longer compete with mass produced small cars of the era, such as the Austin Ruby and Morris Eight, and chose to build more expensive models like the sporting Southern Cross and the larger Gloria saloons, including the elegant Gloria Six Vitesse. Frank had a maroon example of this model as his personal transport. A silhouette of this saloon was displayed on the advertising hoarding affixed to the nearby railway bridge in Walsworth Road (see back cover). On the other side of the bridge, as you entered Hitchin from Walsworth, a further hoarding gave directions 'Keep Left For London SANDERS GARAGE 200 YARDS AHEAD' in large letters. In fact, it seems that Sanders had the monopoly on advertising hoardings on both sides of the railway bridges in Hitchin and Little Wymondley for more than two decades. The diverse range of Triumph models would have appealed to clients of varying means and their reliability would have encouraged a loyal following - good for business. The good relationship that Sanders had established with the Triumph Motor Company was demonstrated on the death of garage founder, Ralph Sanders, when the Triumph Company was represented by Major Hanna, a Director, at his funeral in August 1933. In this later period Sanders were able to satisfy customers who required smaller and cheaper cars by continuing as agents for a variety of manufacturers, such as Triumph's former competitors Austin and Morris.

However, the struggle to keep the business afloat during these difficult times had started to affect Frank's health, later made worse by the devastation of another world war. Sadly, Frank's first wife, Alice, had died in 1931 but just over a year later, in 1933, Frank married Florence Upchurch, a lady some twenty years his junior; a cause for consternation among some members of the family. In 1936 Florence gave birth to Stuart, who first brought the Sanders' archive to my attention. Stuart did not join the family business. He became a chartered accountant, becoming a partner of Bradshaw Johnson, and had little contact with Sanders garage business except when visiting with his father as a child.

Personnel

It has been mentioned that four of Ralph's ten children played a major role in the running of the business but, although what part they played is fairly certain, information about their personalities and personal life has been scarce. It is fortunate that, as they all lived beyond the Second World War, some current members of the family have their own personal memories of these significant characters. Firstly, Stuart Sanders recalls his memories of his father, Frank Sanders.

Ralph Francis Wilkins Sanders

"On 31st March 1897, after leaving the Perse School at Cambridge, my father, known as Frank, was apprenticed to Odell's coach and carriage business in Bridge Street, Hitchin. This business was taken over by his father, Ralph Erskine Sanders, in 1899 as an addition to the existing business carried on at Royston. Prior to joining the business my father was brought up in the family home in Royston but moved to Hitchin after joining Odell's. My father was the oldest of four sons of Ralph and Rhoda Sanders and over the years, up to his father's death in 1933, he assisted in the development of the family business which progressed from crafting horse drawn and other carriages, and building the bodywork of early motor cars, to becoming a substantial garage. My father married Alice Armstrong with whom he had a son, Ralph Bernard (known as Bernard), born in 1905, and two daughters, Betty, born in 1911, and Mary, born in 1920. Alice died in 1931 and, in 1933, my father married my mother, Florence Upchurch; I was born in 1936. In 1933 Ralph, my grandfather, died and my father, who was two weeks off his 54th birthday, became the leading member of the family to run the business. However, he would have already been much involved in assisting his father who had many other business interests, including farming and the purchase of timber, and who by this time had become quite elderly. Apart from the responsibility of the motor business my father was quite involved in the local community and this included being both a Hitchin Urban District and Hertfordshire County Councillor; Chairman of the Board of Governors of St Mary's School; a Governor of Hitchin Boys' Grammar School; a founder member of

Ralph Francis Wilkins Sanders, known as Frank, c.1920.

Frank Sanders with his son, Stuart, from his second marriage to Florence Upchurch in 1937

Hitchin Rotary Club; as a freemason, a Worshipful Master of the Cecil Lodge; a member of Hitchin Bowls Club; and involved with both the Hitchin Mutual and North Herts Building Societies. My grandfather had founded his business in Royston, in 1875, and was in overall charge up to the time of his death in 1933, a total of some 58 years. Although my father took on ever increasing responsibility, particularly

The painting of Kershaw's coach, by W J Shayer, purchased by Frank Sanders for £155 in 1927 and presented to HUDC for the new Hitchin Museum in 1938. (Hitchin Museum)

as my grandfather became elderly, he really only took on the main role in 1933 and, as he died in 1949, he was only the principal for some 16 years. These years were mostly difficult for trade as the Great Depression, which had started in 1929, had recovered very little by 1933 and, indeed, had still some years to run. In September 1939 the Second World War broke out and, with all the problems of rationing and staff being called up, the business was very much restricted. After the war ended in 1945 there was a period of austerity which lasted until well after my father's death. So, in short, there were only a few years just before the war when the business environment was helpful to trade. Also, my father became increasingly ill, particularly during the last two years of his life.

The following are some of my memories of my father but, as I was only 12 years old when he died, they are somewhat limited. My father did not discuss business matters with me but I often spent time at the garage – probably making a nuisance of myself. There was a lift on which vehicles could be raised to the first floor and this was operated by a rope on a differential system so that vehicles could be lifted by manpower. There was a large loop on this rope when the lift was not in use. This loop made a wonderful swing which I used much to my father's annoyance – presumably he considered it was not a safe thing to do. On the first floor was stored the original stage coach used by Kershaw's to provide a service between Hitchin and London up until 1850 when the railway was too great a competition. Sometimes, with a school friend, we gave each other rides by pushing it along a few yards. On 2nd July 1927 my father purchased at Sotheby's an oil painting of this coach by the artist W J Shayer for the sum of £155. He presented it to the Hitchin Museum at the Town Council meeting on 21st December 1938. After Hitchin UDC was absorbed by North Herts District Council the painting was moved to the new council offices at Letchworth. I felt that this should not have happened and, with the help of Councillor Mrs Beryl Wearmouth, managed to have it returned to the museum in Hitchin where it has been prominently displayed. The coach was restored for the Hitchin Pageant of 1951 held at The Priory. My father drove a maroon Triumph Gloria Vitesse motor car and, as a passenger, I often stood on the front passenger seat without any support which, of course, today would be considered very dangerous but fortunately there were no unpleasant incidents. I do, however, remember the leather seat becoming quite worn. I remember hearing about my father's role in the First World War when he was a special police constable. He was posted, together with truncheon and whistle (both of which I still have as souvenirs), to Benslow Lane outside the German Hospital, as it then was, as this building overlooked the Great Northern Railway whose main line passed by in a deep cutting. So presumably he was there to protect the line in case the hospital's occupants were intent on using the location to sabotage the track or trains. I remember the County Council election in 1946 when my father was elected as councillor for Hitchin South. He stood as

an independent as he believed national politics should not necessarily influence local issues. As a councillor I remember going with him to Hertford where he had a meeting expected to last two hours which gave me time to explore the town. The meeting actually lasted only 10 minutes but no one knew just where I was until I returned to County Hall nearly two hours later. I also remember, during the war years, evacuees from Eastbourne and some female service personnel being billeted on us and of enemy aircraft flying overhead. Part of the time we sheltered under a steel table in the kitchen but later we used the cellar. I can just remember the night of the infamous raid on Coventry when we spent pretty well all the night in the cellar and I cheerfully asked "When are we going to die?" I also remember seeing the glow from the tails of the V1 rockets passing over Hitchin."

Next, Joanna Cooper (née Sanders), Roy's daughter, who gave information about the Central Wireless Stores in the previous chapter, now recalls her memories of her father and also those of Lance and Doris:

Royston Jonathan Sanders

"My father, known as Roy, was the youngest of the ten children - four boys and six girls - of Ralph & Rhoda Sanders. He was born on August 7th 1893 in Royston (Cambridgeshire, as it was then, but after a boundary change it was moved into the jurisdiction of Hertfordshire). The family lived in York Villa in Kneesworth Street,

Jonathan Royston Sanders,
known as Roy, c.1950

Royston, a tall terraced house, and later moved to a large Victorian semi-detached house in Kneesworth Street with an extension built on the side to accommodate the large family. I know very little about my father's childhood except for family anecdotes. He was very fond of his Aunt Lizzie (Cromwell) who lived in Buntingford and he often stayed there. This is where he learnt to swim in the ford leading up to the church in Layston. Aunt Lizzie ran her own dressmaking business assisted by her niece. She must have changed course later and ran a cake shop in a little house in the High Street on the corner with Cottered Road. My father took me to see it. It had a little flight of steps up to the front door. As a small boy he used to attend chapel with the family who always sat in the same pew. During the sermon, my father's attention was drawn to a large lady sitting in the front of him wearing an extravagant bonnet with flowers and bees on little wires. As she moved so did the bees which rocked to and fro and kept him amused. Also connected with chapel service, he used to make up his own words to well known hymns which even now I am unable to sing without laughing. One of his amazing talents was that, without any musical tuition, he taught himself to play the piano and could improvise and play any song of the time without music – but always in the key of E♭ I can see him now playing anything I suggested in complete confidence. One of his sisters was also an accomplished pianist. He, like his older brothers, went to the Perse in Cambridge. I have the impression he left school at 14, or soon after, and went into the family

Captain Jonathan Royston
Sanders c.1914

business, certainly not his choice. He would have loved to have continued his education in science or physics; he was especially fascinated by electricity. With the onset of the First World War he, with his older brother William, volunteered and joined the Leicestershire Regiment. Their mother Rhoda's (née Wilkins) family came from that county and they probably wanted to keep up that connection. Rhoda's father, Jonathan, may well have provided my father with his Christian name. My father was sent to Egypt and served in the Western Desert Campaign in the Machine Gun Corps incorporated with armoured cars which were the first motorised vehicles to carry machine guns. These armoured cars, introduced in 1915, were known as LAMBs (Light Armoured Motor Battery) and were built on the Silver Ghost chassis by Rolls-Royce with four wheels at the rear to take the additional weight. They had a turret on top fitted with machine guns. To protect these vehicles from enemy gunfire boiler plates were riveted onto the body on all sides. This not only made them very heavy but they could be lethal because, if hit by a shell, they would shatter sending shards of steel flying about the interior. With his practical knowledge of cars and car engines he was among a very few who could drive these vehicles. After the war, my father wrote to Rolls-Royce praising them for the quality and

Letter sent to 'Roy' Sanders in reply to his praise of their armoured cars, in which he served during the First World War, by the London manager of Rolls-Royce Ltd., dated 6th January 1919.

versatility of their vehicles in such extreme circumstances. In reply he received a letter from the London Manager of Rolls-Royce who said that they were "glad to hear that our cars have given every satisfaction under difficult conditions". My father returned to the business and, in 1924, married Elizabeth (Christine Frances) Goodman, also from Royston, and started married life in the three storey Victorian house adjacent to the garage in Walsworth Road. My brother John was born in 1927 and for his 4th birthday my father asked him if he would like a car for a present. On an excited affirmative reply, he said to him "let's go round and you can choose one". Hand in hand they walked to the showroom (virtually next door) and positioned within a semi circle of some magnificent and very highly polished motor cars was a bright red pedal car, complete with proper number plates. Without hesitation my brother said "that one" & drove it out of the showroom as the proud new owner. In 1953, by which time we had moved to live in Wymondley Road, and just one week before my brother's wedding, my father who loved working out surprises, gave him & his wife-to-be a small ball of string and asked them to wind in the string and their wedding present would be at the end of it. They walked miles backwards and forwards round the garden and eventually they came to the end of the string which was tied to the door handle of a Morris Minor parked in the lane at the end of the garden. My father loved doing this sort of thing. Another incident that my father told me was when Sanders were servicing an Austin 7 coupé belonging to Miss Chambers,

headmistress of the Girls' Grammar School. Admiring the vehicle, now ready for collection, he was eating an apple. He was obviously interrupted in his indulgence and hastily put the apple core in the folded hood of the car and forgot about it. Next day a large wet stain had appeared resulting in the need for a brand new replacement hood – an expensive and unpopular action! He had a good sense of humour, for the most part harmless and based on names, words and little songs. We as a family understood them, however repetitive they were, although it could be embarrassing when sometimes used in the company of complete strangers; more than once I had to extricate him and me with the excuse that it was a family joke! I think he would have liked to have been a magician too and he was always amused by Tommy Cooper. He would spend hours soldering and making tricks in his workshop but it was sad that so few of them worked! This sense of humour, I'm sure, stood him in very good stead when once again he volunteered to serve in the Second World War. He would have been 41. For most of the war he was stationed in Carlisle. He joined the Royal Army Service Corps (RASC) in charge of the Drivers' Training Battalion where thousands of young soldiers learnt to drive tanks, lorries and all manner of army vehicles, including driving convoys at night with, of course, no lights or sign posts. After the end of the war my father returned to the business. Like so many others he found it difficult at first to adjust to civilian life. So much had happened in those six years. My brother was 12, in 1939, when he started at the

Hitchin Boys' Grammar School and in 1945 he went up to Cambridge University resulting in a PhD in Natural Sciences, thus my father missed out on his son's formative years, a time that would have interested him so much. His brother Frank suffered ill health during this period, probably caused by the strain of keeping the business going during the war with petrol rationing and other restrictions. There was much to do, including creating more workshops for the repair of vehicles, particularly commercial vehicles used for war work. However, my father took a great interest in the town. He served on Hitchin Urban District Council for many years as an independent member. He was Chairman of North Herts Building Society; a Rotarian; and an active member in Hitchin Chamber of Commerce along with many well known Hitchin names. He also organised a Reunion, always on Cup Final Night, of his RASC Battalion in the Shakespeare Inn in London. He was a family man and loved getting us all together. I know we all remember with great pleasure (not always with success) croquet in the garden during the summer. Of course, Christmas too, especially sitting around a roaring log fire with my father puffing a cigar and reading 'Tales from Pickwick Papers' to us. He loved his garden and he and my mother, who was so creative, designed a new garden when they moved house. He was a great 'potterer' and when not in the garden or having a '5 minuter', as he called an after lunch snooze (more usually ½ an hour), he always kept himself busy." Roy retired in 1960 and died in November 1980, aged 87.

Roy, like his three brothers, had once lived in the house at 58 Walsworth Road, immediately next to the garage. In 1935 he had a new house built at 14 Wymondley Road which he called 'Layston', evoking memories of his childhood home at Layston Park, Royston. In the 1960s he had another house built in the garden to the side of his former home, opposite the top of The Avenue, on land originally earmarked for the continuation of The Avenue through to St John's Road. He took the name 'Layston' with him to his new home. Shortly after, with one other owner in between, the original house was purchased by Stuart Sanders, Frank's son, who renamed it 'Jesmond' after the area in Newcastle-upon-Tyne where his mother's sister and her husband, the eminent gynaecologist Professor E Farquhar Murray, had lived and where Stuart regularly visited as a young man.

Here are Joanna Cooper's memories of her uncle, Lance Sanders.

Launcelot Vivian Erskine Sanders

"Launcelot Sanders, known to his nieces and nephews as 'Uncle Lanny', was the second son of Ralph and Rhoda Sanders, born in Kneesworth Street, Royston on 8th May 1881. Like his older brother he was educated at the Perse School in Cambridge. When he left school, or even before, he may well have accompanied his father to farms in a wide area in Hertfordshire and Essex with a view to purchasing

Launcelot Vivian Erskine Sanders, known as Lance, in 1927.

timber for the coachbuilding business. This introduction would have appealed to Lance being the most practical member of the family. He joined the business and probably did an apprenticeship in carpentry and cabinet making as precision in panelling, wheels, shafts etc. was paramount in the construction of these beautiful coachbuilt carriages and later motor cars. Lance was a tall handsome man with a handlebar moustache and appeared in many of the photographs of the business. He married a farmer's daughter from Clavering in Essex, Daisy Custerson, a very well read lady, a Quaker and very interesting to talk to. Lance was a man of the land, a keen gardener and kept chickens, keeping our family well supplied with eggs, always beautifully packed in straw in lovely wooden boxes with sliding lids and dovetailed corners – part of his expertise. He lived at 6 Wymondley Road and, during the war, with petrol almost in non supply, he could be seen taking his car out of his drive way and, with the car door open, would 'leg' the car into motion on the slight decline into The Avenue and then freewheel down the hill which gave the car enough momentum to swing into Walsworth Road and to reach the garage forecourt. He owned an Austin 10 fitted with a wooden dashboard which had the latest accessory attached – a chrome vase in which he always placed a fresh flower every morning. Lance and Daisy had no children of their own but always kept sweets and cigarette cards for when my brother and I went around to see him. Lance was a bit of an eccentric and, in later life, had a pair of wing mirrors from a Morris Minor mounted either side of his bath so that he could shave whilst bathing." Launcelot died in 1970.

Lastly, Joanna Cooper recalls her memories of her aunt, Doris Sanders.

Doris Miriam Sanders

"Doris Sanders, known to her nieces and nephews as 'Aunt Doll', was born on 5th March 1888 – a surviving twin – the seventh of the 10 children of Ralph and Rhoda Sanders. She and her sister, Irene Mary, 16 months younger, lived at Layston Cottage in Royston, one of the lodges to Layston Park, the family home. Doris was the only daughter who had an active part in the family business. She was responsible for the day-to-day accounts, wages etc. I remember

Doris Miriam Sanders, c.1910

watching her add up columns of figures by running her pen up the page at great speed and, of course, accuracy, urging me on to do the same – no contest! Aunt Doll, like three of her sisters, never married but was very fond of children. She used to play with my brother and I when we were small and made up stories as she went along; she was always interested in us. Travelling to work from Royston to Hitchin became impossible by car during the war so she stayed in Hitchin during the week and went home at weekends. She had a 'bed-sit' in a house in Chiltern Road just around the corner from us and, every Tuesday, would have her lunch hour with us before returning to the office in Walsworth Road." Doris died in 1956.

William Humphrey

Bill Humphrey (right) with Bill Blows, the blacksmith, c.1950.

In 1933, Frank gave employment to his cousin, William Humphrey, known as Bill, another example of a member of the family being offered a job. Fortunately, in 1997, the Society recorded Bill's son David (in connection with another publication) in which he shares his memories of his own and his father's involvement with the company. David was born in 1930.

David Robert Humphrey

David recalls *"My mother's family had farmed near Oakham, Rutland for generations but it was not the life that appealed to my father. His interest was engineering and he became an apprentice with the Pick Motor Company in Stamford. I remember him saying that, if he didn't get to work by 6.00 a.m., he was locked out until after lunch and received no pay! After serving his time he went to Luton one Saturday with some friends and, whilst loitering outside the Vauxhall factory (before the purchase by General Motors), they encountered a gentleman who asked them why they were there; they said that they wanted jobs. It turned out that he was a Vauxhall director and he told them to report for work at 8.00 a.m. on Monday morning. During the First World War my father served in the Royal Naval Air Service flying in sea planes. Later, he returned to the farm, still worked by his elder brother, but due to the combined effects of the Great Depression, and problems with farming generally, there was no work for him. It was then that his cousin, Frank Sanders, offered him a job in Hitchin. Firstly, he travelled down on his own and stayed at the Hill View Café in Bridge Street, run by Mrs Bottoms; my mother, my brother Donald and me joined him a few months later when we moved into 58 Walsworth Road, immediately next door to the garage.*

Several members of the Sanders family had lived in this house over the years; it had been a recruiting office during the First World War. My father became the workshop manager, replacing Mrs Bottoms' husband who previously held that position, and stayed there until his premature death, at the age of 64, suffering ill health in his later years. During the war I helped my father in the garage after school, often working late into the night. Sanders Garage was the place where any aspiring motor engineer wanted to start their career. Local men who started businesses in Hitchin included Jack Maidment, who founded Maidment's Coaches in Victoria Road, and Bill and Cecil Sale who later ran a garage in Bridge Street close to the river, notable because the building fell down one day without warning. The garage was a busy place and I remember Bill Hall, the blacksmith, making springs; batteries were also re-plated. I learned to drive when I was six and, during the war, drove my own car, an Austin 10 4-seat tourer, always with the roof up. I remember one customer, Maurice Walker, who worked for the seed merchants Henry Franklin in Biggleswade. At the beginning of the war he bought two new cars from Uncle Frank, a Morris 14 and a 2-door Morris 8. I had to change the clutch on the Morris 8 every six weeks. Walker was a very big man, probably 19 stones, and there was nowhere for his left foot so he rested it on the clutch pedal, eventually burning out the clutch. Every second visit for this job I had to repair the seat frame which regularly broke. I dismantled the seat and my father welded the damaged tubular frame. I also remember, early in the war, the ambulances made by Cecil Saunders of Letchworth. They cut the bodies off big American V8 saloons like Packards and Oldsmobiles, from behind the driver's door, and fitted tongue and groove wooden platform bodies with canvas tilts. We made and fitted much stronger road springs to cope with the extra weight. During the war many garages closed but Sanders remained open. There were a row of six petrol pumps outside with swing arms over the pavement to reach the cars. They, and George King's factory next door, were guarded by two armed policemen with fixed bayonets, although I suspect they never had any ammunition. The pumps were allocated to different organisations; I think that they were the army, the fire brigade, doctors and nurses, the grey ambulances, the London & North Eastern Railway and one for us. I did enjoy my school days at St Mary's School, although Mr Spikesman was a very strict disciplinarian. He was particularly strict about polished shoes and clean finger nails; I struggled most with keeping my nails clean despite using bleach. Whether it was my untidiness, or the fact that, at age 15½, I was smoking 20 cigarettes a day, or for some other reason I don't know but, in 1945, I was asked to leave school - another interpretation could be that I was thrown out! When I told my father he was understanding but told me that I must tell Uncle Frank. First he offered to take up my reinstatement with the headmaster, Uncle Frank was Chairman of the Board of Governors so had some 'clout', but, when I said I didn't want to go back, he simply said that I had better come and work at the garage. My father wouldn't agree. He said that I should do it properly and get an apprenticeship like he did. So Uncle Frank arranged for me to go to Rootes in Coventry. They agreed to take me on, but for a five year apprenticeship rather than the seven years that my father had served, although I had first to do six months on probation as I was too young for my father to sign the indenture. On my visits home I often chauffeured Uncle Frank in his Triumph Gloria when he became too ill to drive".

David's post-war exploits, eventually returning to Hitchin to work for Sanders, are given in the next chapter.

Donald Humphrey

Doris took Donald Humphrey, Bill Humphrey's elder son, under her wing and trained him to deal with the company accounts. Donald had been run over by a car as a child, family legend says by a Rolls-Royce, which resulted in him being *'a little slow'*. When Doris suffered ill health in later life Donald was able to take her place. He continued to work in Sanders' accounts department, and later for the Stevenage Motor Company, until his sudden death aged 59.

Jubilee Exhibition 1935

The Sanders' archive contains a local newspaper cutting (name of paper unknown) from 1935, the Jubilee Year of King George V, which reports on an exhibition held at Sanders garage in Walsworth Road, on both the ground and first floors, where a variety of vehicles, both old and current models, were exhibited. In addition to an early boneshaker bicycle there was an 1894 'Cannstatt Daimler', so described to explain that it was a German product, built in Bad Cannstatt outside Stuttgart, rather than a British Daimler. There is a photograph of the car with Roy in the driving seat and Frank on the rear seat. It was registered AR 2, only the second number issued by Hertfordshire County Council. The report says *'It seems almost unbelievable, in the light of present day values, that this car involved an outlay of something like £800, for today one can buy the most princely of cars at a figure much less than £1,000.'* However, the thrust of the report continues *'The average man … seeks for something well within his means … and very often running costs, taxation and insurance are the main considerations which weigh with him when investing in a new car.'* The cars on display included models made by Talbot, Crossley, Triumph, Wolseley, Humber, Austin and Morris. They were offered for sale at prices ranging from £120 for a Morris 8 to £850 for a Talbot 20 and demonstration runs were available. Commercial vehicles were displayed alongside the cars. You could also visit the Murphy Radio stand and have a cup of tea from the light refreshments buffet. The report concludes *'An hour at this exhibition is an hour well spent'*.

Motoring Regulations

The Road Traffic Act of 1930 abolished all national speed limits and introduced compulsory third party insurance. The Road Traffic Act of 1934 introduced the driving test - with effect from 1st June 1935 all those who had been driving since 1st April 1934 had to pass a compulsory test. This act also reintroduced a speed limit for cars of 30mph in built-up areas but there was no national speed limit again until the 70 mph limit was introduced in 1965.

The Second World War

The business had survived the lean years of the thirties very well, helped by the popular Triumph agency and the company's ability to carry out any type of maintenance or repair to a motor vehicle made possible by the wide ranging skills of its staff and its comprehensively equipped workshops. Frank had the support of his new wife who had given birth to a son, Stuart, in 1936. However, as mentioned by Joanna Cooper, the strain of managing the business was beginning to have an effect on his health and he began to rely more on his son Bernard. If that wasn't enough, the declaration of war in 1939 had a major effect on the viability of the business.

In the country as a whole, the introduction of new regulations by the government was to transform the lot of the motorist for more than ten years, long after the war had ended. Firstly, petrol rationing was introduced, on 23rd September 1939, with the intention of the basic ration restricting motoring to about 200 miles per month; the number of coupons being based on the car's horsepower. Early in the war the anonymous 'pool petrol' (so called because the petrol companies pooled their resources) was readily available and cost 1/6d per gallon but, by 1942, it had risen to 2/2d. David Humphrey remembered that, at Sanders' Garage, different organisations had a specific pump allocated to them for their supply. Supplementary coupons were available for essential work but, in July 1942, the basic ration was withdrawn completely leaving only those that could prove their journeys were essential getting any petrol coupons at all. Motorists also had to cope with the blackout with the use of only one headlamp fitted with a mask having a narrow slit or slits in it and the edges of the car painted white. Whether Sanders sold these masks or, indeed, the white paint, is not recorded. The limited import of rubber from the Far East, due to Japanese conquests, affected the availability of tyres in the UK and they were restricted to those with 'E' (essential) petrol coupons. In July 1940 car manufacture came under government control and there was only limited production of basic models from Austin, Morris, Ford and Hillman intended for motorists on essential work, although the new purchase tax had, in any event, made them expensive to buy. Lack of resources of all kinds was not restricted to materials as Sanders' workforce was also depleted as employees were called up to fight for their country.

David Humphrey mentioned that many garages closed during the war but Sanders carried on. Neil Thomson, whose father ran a garage in nearby Queen Street, told me that this nearly didn't happen because, early on in the war, the army arrived in the town along the A505 from Letchworth and originally intended to requisition Sanders' Garage in Walsworth Road as a repair shop for army vehicles. Frank, having some influence in the town, told them that his other garage in Bridge Street would be more appropriate so they continued along Walsworth Road into Queen Street. However, before they reached Bridge Street they came to Thomson's Garage and chose to requisition those premises instead. Neil's mother was told to move out, her husband 'Jock' Thomson was away in the RAF, and all their vehicles were moved onto St Mary's Square - a close shave for Sanders!

During the war many motorists were forced to give up driving and either sold their cars or laid them up for the duration. Others 'downsized' to a more economic model. That was the choice made by one of Sanders' customers, Sir Ralph Delmé-Radcliffe of Hitchin Priory. His Rolls-Royce Silver Ghost, fitted with a 2-door coupé body by Harrison, originally purchased for his wife, was exchanged at Sanders' Garage for an Armstrong Siddeley saloon. The garage made use of the Rolls-Royce by removing the boot lid and fitting a crane for use as a tow car. After the war the car was purchased by the well known local vintage car enthusiast and dealer, Cecil Bendall, for £50, who then sold it at auction. Bernard agreed to the sale as long as he could keep the mascot, now used by his son Robert as a door stop! It is believed that the car spent time in the USA and Switzerland but is now back in this country, although its current whereabouts are unknown.

With the range of Sanders' services being reduced during the war years, part of the upstairs of the garage was let to Geo W King Ltd whose engineering business, involved in war work, was immediately next to Sanders' Garage, in fact a first floor entrance was created between the two buildings. They vacated the garage when their business expanded to a new factory in Stevenage in 1952. Herbert Sharpe, the banner maker, moved his studio into another part of the first floor (towards the town) after he was bombed out of London. He painted the beautiful silk banners carried by trade unions all over the country, as well as making sashes for beauty queens and flags. He passed the business to his daughter who relocated it to Tilehouse Street in 1957.

The combination of all these obstacles resulted in less business for Sanders during and immediately after the war, although they are believed to have survived better than many of their competitors. However, by the time that hostilities had ceased Frank was 66 years of age and the stress of those difficult years had impacted on his physical health. Despite this he stood for election as an independent candidate and was elected as a county councillor, Hitchin District, in 1946. Frank died in 1949.

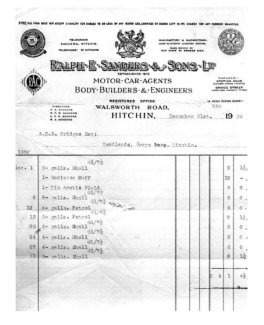

Two bills issued. (Left) To Mr A B N Bridges of Grays Lane, Hitchin for 35 gallons of petrol, Arctic Fluid and a Radiator Muff totalling £4.1s.4½d., dated 31st December 1932. (Right) To Mr H Clark of Redhill Road, Hitchin for two 4.50 x 17 Firestone covers (tyres) and one Champion L10 spark plug, plus an outstanding account, totalling £8.0s.10d., dated 31st December 1943.

Post War Recovery:
new hands at the helm

This chapter explains how the business recovered following the Second World War with new hands at the helm. As there is very little documentary evidence for this period I hope that the memories recorded by those involved in the business, often very amusing, will help to tell what was going on during this time and bring Sanders Garage to life.

Shareholders

Following the death of Frank Sanders, the company continued in family ownership into the 1950s, with the dividends from many of the shares providing an income for the maiden aunts. However, on 15th December 1954, the company share register records that a substantial interest in the company was sold to the Stevenage Motor Company. They purchased 4,125 shares from various members of the family. Also at this time two directors of the Stevenage Motor Company, Major Francis Grimsdick Clarkson JP, a military rank that he achieved during the Second World War, of Townsend Close, Stevenage and Francis Charles Layfield Broadribb, a farmer who had been employed by Morris Motors, of 4 Julians Road, Stevenage, also acquired shares personally - Clarkson 40 shares and Broadribb 20 shares, all previously owned by Roy Sanders. In 1957 Stuart Sanders received 500 preference shares and in 1959 David Humphrey received 100 ordinary shares.

Walsworth Road

Immediately post war the business concentrated upon the sale, repair and maintenance of second hand cars and light commercial vehicles. Later, the sale of new cars re-commenced, although there were few available to sell as materials were in short supply and government regulations gave priority to specific types of customer in the UK, such as doctors, and the majority of car production was exported to earn much needed foreign revenue for the country. With the Triumph Motor Company going into receivership in 1939 it was necessary for Sanders to approach new manufacturers to supply the garage with cars. In what sequence is not known but by the 1950s they were agents, and later dealers, for Rover and Jaguar and also Morris, Wolseley, MG and Riley which, in 1952, as the Nuffield Organisation, merged with the Austin Motor Co. to form the British Motor Corporation (BMC). In 1966 BMC took control of Jaguar and, in 1968, merged with the Leyland Motor Corporation Ltd., who had already purchased Rover in 1967, to become the British Leyland Motor Corporation (BL). These are the makes of cars that Sanders sold throughout this period until the garage closed in 1979. The models included many favourites like the Morris Minor, Mini and Oxford at the lower end of the market, to the 'Cyclops' Rover 75, and later 2000, and the Jaguar Mk2, and later XJ6, for more affluent customers. Light commercial

(Left) The Walsworth Road showroom, illuminated at night, in the 1950s, with a Ford Popular parked outside and the petrol pumps still in place. (Right) The Walsworth Road garage, viewed from the petrol station, in the 1960s, with a Mark 1 Jaguar parked outside and a Hillman Minx in the street. (Below left) The Walsworth Road showroom in the 1960s. Various models of car for sale can be seen through the windows. (Below right) The petrol station in the 1960s, with a Morris Oxford filling up. The lubrication and wash bays can be seen to the left of Desmond House.

vehicles made by BMC, and later BL, were also sold like the popular Morris Commercial J-type, J4 and later Sherpa, vans. There are photographs elsewhere in the book showing some of these vehicles in or outside the showroom, together with an advertisement from 1977.

Personnel

Frank Sanders having died in 1949, the daily management of the garage was taken over by his son, Bernard, who had by then worked in the business for over 25 years. Roy continued as an active member of the management team and became Chairman of the company. Lance continued to assist until his retirement in the 1950s but in a management rather than a manufacturing role. Another member of the management team in this period was Mike Anderson who had been born and grew up close by at Walsworth House. On the death of his father, Major F H Anderson, in 1953, the house became a seminary for boys who wished to become Catholic priests and, in the 1960s, was purchased by Hertfordshire County Council and forms part of North Herts College.

Firstly, we hear about Bernard from his son, Robert.

Ralph Bernard Sanders

"Ralph Bernard Sanders, known as Bernard, was born on the 23rd April 1905 at 78 Walsworth Road, Hitchin and was the first child of Ralph Francis Sanders and Alice Emily Sanders, formerly of Hillside Farm, Preston, near Hitchin. Two further children were to follow, Betty and Mary. Dad went to school at what is now the Girls' School in Highbury Road which then taught both boys and girls (in the kindergarten). He later went to Oakham Public School and left to join the business in 1921 at the age of 16 years. With the business prospering his father, Frank, bought 'Burley', a much more prestigious house in Highbury

Ralph Bernard Sanders, known as 'Bernard', c.1970.

Road. Dad married my mother, Evelyn May Wallace from Walsworth Farm, in 1930 and set up home at 76 Walsworth Road, later to move to 'Bridlemere' also in Highbury Road in 1938. Bernard worked his entire life at the garage and recalled many of the aristocratic customers who purchased early cars which the company bought as chassis to be fitted with quality bodies and interiors. He used to reminisce about some of the customers of those days, one in particular was a Miss Cotton Brown of Walkern Hall, a rather unaccommodating lady who employed a chauffeur, Mr Adams, who sometimes cheated upon keeping the car clean by only cleaning the side that 'madame' regularly *got in. Another customer was Sir James Hill, and a Mr Pounsby from Caldicot Lodge which still stands in a forlorn state by the side of the A1 with a travelling burger bar at its frontage in a lay-by. During the war he was involved in the ARP and talked about various members of this service including Rene Sharp and Queenie Rix. I remember well after the war, when rations were still in operation, dad came home with tins of Smith Crisps and cakes made by a Mr Swannell who, I think, had a shop in Bancroft, probably as 'rewards' for petrol. Dad was very much a 'hands on' man; I remember him checking car batteries being charged in the cellars and making up car registration number plates when there were staff shortages. He always complained about pricing up workshop repairs for invoicing to customers. He also hated manipulating the telephone exchange, particularly when more than one line started ringing. We were an AA garage and were obliged to attend breakdowns and accidents, even in the middle of the night. He and Bill Humphrey, the caretaker who lived next door at 58 Walsworth Road, used to go out and assist using an old open top maroon Rolls-Royce because it was so powerful to pull cars out of ditches - later to be replaced by a Land Rover. I remember going with him, and Charles Cushing the workshop foreman, to Kings Walden where a Miss Medlicott lived and collected a Morris Minor Traveller which was under performing. The lack of performance was a complete mystery. After several attempts to correct the fault without success, Miss Medlicott managed to converse with Lord Nuffield (founder*

Bernard with his wife, May, and their two children, Margaret and Robert, on their Silver Wedding Anniversary in 1955.

of Morris Motors) airing her dissatisfaction but it had the desired effect which was traced to a faulty camshaft profile. (Dad used to say 'oh no tell her I am out' when she called!!) Dad retired in 1972 but still enjoyed collecting and delivering cars for the business which was, by now, part of a bigger concern. He died in January 1990 at the age of 86".

Another person who enjoyed delivering cars in the later years of the business was **Betty Detmer**, whose husband was a director of Hawker Siddeley in Hatfield.

David Robert Humphrey

David's reminiscences of his early years, leading up to his taking an apprenticeship with the Rootes Group in Coventry, in 1945, were given in the previous chapter. He now relates his post war career, including his joining Sanders to take over as manager in 1958.

"I continued my apprenticeship with Rootes receiving training in many departments but I was destined to become a service representative. By the time I was 23 years old I was 'service rep.' for Scotland and living in Edinburgh. I made many useful contacts in the motor industry during this time which came in very useful later when I returned to Hitchin. When I was 19 years old I entered my first car rally. My first Monte Carlo Rally was as a private entrant in a Hillman Minx but I later co-drove for the successful Rootes works

David Humphrey, Leslie Johnson, Stirling Moss and Johnny Cutts, all seated on the bonnet of a Humber Super Snipe, who took part in a publicity run in 1952 from Oslo to Lisbon taking in '15 countries in 5 days'. (The Sunbeam Talbot Alpine Register)

Peter Harper and David Humphrey standing beside their Sunbeam Talbot MkIII in which they came 9th overall in the Monte Carlo Rally of 1955. (Nic Cooper)

team with Stirling Moss, Sheila Van Damm, whose father managed the Windmill Theatre in London, and local man Peter Harper, who ran the Rootes Group agency in nearby Stevenage. We competed in all the big rallies in the 1950s in Sunbeam Talbots, Alpines and Rapiers, including the Tulip and Alpine Rallies and a total of ten Monte Carlo Rallies. In 1952, I was part of a Rootes publicity stunt with Stirling Moss. We had to take a Humber Super Snipe from Oslo to Lisbon, passing through fifteen countries in 5 days; we actually did it in four days! In 1955, I finished 9th overall in the Monte Carlo with Peter in a Sunbeam Talbot MkIII and in 1957 I did the Mille Miglia with Sheila in a Sunbeam Rapier but we crashed into a shop window and didn't finish. I regularly visited my parents in Hitchin and during one visit, in 1958, I was asked by Major Clarkson, who owned Stevenage Motor Company and, by then, also a large holding in Sanders, to join them as manager of the Hitchin garage. Rootes had offered me a better paid job in California but I was fed up with travelling. Clarkson said that, if I worked hard, I could be on the board in a couple of years. I did work hard and was made a director in nine months. The house next door to the garage, at 58 Walsworth Road, was converted into two flats with my wife, Janet, and me in the ground floor flat and my mother in the upper flat. With my involvement in motor sport I made friends with Raymond Baxter and Kenneth Allsop [both BBC TV presenters] and they became customers. Contract hire was one of our core operations with Randall's of Bedford one of our larger clients. Around 1960 we built a big new workshop behind Desmond House and a petrol station in front. We sold and serviced Rovers and Jaguars and the full range of BMC cars. We employed mechanics, electricians, paint sprayers and a blacksmith.

The workforce in the 1960s. L to R: Mick Bishop, Ted Challis (stores manager), Colin Orchard, Alan Primett, Charlie Cushing (white coat), Ken Smith (above), Alan Burling, Ray Hart, Vic Grimwood, Paul Presland, Lincoln Ansell, Geoff Clark and William Woods. (Geoff Clark)

In 1962 Stevenage Motor Company bought the remaining shares in Sanders from the family and took total control. I became managing director and stayed for several years."

Amongst those employed in the workshops during these years was Ted Burrows, an electrician, and Bill Blows, a blacksmith, together with many more referred to in the following memories of their colleagues. In the early post-war period Sanders recruited several young men to fill new vacancies as the business gradually recovered and expanded. Three of those young men, who started apprenticeships with Sanders during this period, tell their story. Some of their tales are slightly alarming but, no doubt, that's what makes them all the more memorable. The recollections of one of the secretarial staff are also shared.

Geoff Clark

Here are Geoff's reminiscences, known when he worked at Sanders as 'Nobby'. *"I left school at the age of 15 in 1960 and joined Sanders as a trainee motor mechanic, staying there until I left in 1977. I completed my apprenticeship at the age of 21 and was trained how to repair motor cars, including welding. I also learned how to fit new tyres and to repair punctures – there were few tyre depots in those days. I worked from 8.00 a.m. to 6.00 p.m. Fred Coutts was an auto electrician and Paul Presland, known as Algie, was his apprentice. Cecil Cowley from Offley was the blacksmith. He made new leaf springs on the forge. He was helped by Bill Blows, until he left to join Shelvoke and Drewry, and who was later to become my father-in-law; his daughter, Jane, worked in a sweet shop just down the road from Sanders which I frequented from time to time and she explained that her dad had worked at Sanders some years earlier. We were married in 1969. The new workshop was opened behind the petrol station around the time that I started. I remember a Jaguar, newly repaired by Izzards Coachworks, rolling down the sloping forecourt and being damaged in the very same area that had just been repaired. The mechanic's foot had caught the handbrake, positioned on the right of the seat on those models, and accidentally released it. On another occasion a car I was driving was damaged when the brakes failed as I drove it into the workshop – even though I had been told "Watch the brakes, they don't work!" The paint shop was upstairs at the station end of the building. There were fifteen employed in the workshop. We were all issued with an orange seat cover to keep the seats of customers' cars clean. We wore navy overalls which were washed weekly. A frequent car in for repair was Dr Grellet's as he was hard on clutches; he could burn a clutch out in 1,000 miles. I also remember repairing the clutch on the car of local clothier, Leslie Leete. Mr Leete had strange driving habits; on returning to his house with him driving his car I was amazed to watch him pull away in first gear, miss out second and third, and drop straight into top causing the car to struggle. Perhaps that led to the damage to the clutch. Other regulars were the cars owned by the McMurtries from St Ippollitts, Olga Sainsbury, the Spencer Smiths from Highdown, the Ransoms who owned the distillery, estate agent and county councillor Jimmy Hill, Cam Gears and Kenneth Allsop, who had a Jaguar XK140 fitted with hand controls. Most of the cars worked on were the types sold by the dealership, including Morris, Rover and Jaguar. As well as repairing cars we also fitted seat belts, tyres, radios and under-sealed the underside of new cars. A job we all tried to avoid was greasing a customer's invalid carriage, there were about twenty grease nipples to find, and the owner watched and made sure you greased them all. I remember one occasion when a senior mechanic had to do the job because the rest of us had strangely disappeared. The coachworks, next to the new workshops, were run by Ted and Tom Izzard who repaired damaged bodywork and fitted vinyl roofs. Sanders were the first garage in Hitchin to have pumps outside to dispense petrol. They also made car batteries. David Humphrey dealt with car sales and Donald, his brother, was in the accounts department with Isobel Ashpool. Ted Challis and Pete Meadows were in the spares department, Ted was the manager, and Bruce Camfield was a mechanic. Pete and Bruce also organised the last works reunion. I remember an occasion when I was welding underneath a customer's car and the back seat caught fire blackening the headlining. None of the old foam fire extinguishers worked but luckily someone found a modern one that did; just as well as two gallons of foam in the car would have done even more damage.*

The damaged items were replaced with second hand spares bought from Jack's Hill scrap yard. Another fire occurred when an engine 'flooded' with fuel; the plugs had been taken out and the engine was turned over on the key causing petrol to shoot out which was ignited by the plug leads. Fred, who had caused this to happen, panicked and ran out of the workshop, so I slammed the bonnet shut to put out the fire! An old 500 gallon petrol tank, in front of the car showroom, was where the waste oil was stored and then pumped out and recycled. I remember we used to have races around the workshop on trolley jacks. We also built a go kart when times were slack. In the service bay we had grease guns on the end of long reels and we used to shoot each other with the grease or target apples on the trees in the garden. I spent a happy seventeen years there."

John Brown

Here are John's reminiscences. *"I worked at Sanders garage from 1946 until 1975. My mother took me to see Frank Sanders for an interview. I still have my indentures which record my first wages – £1.0s.9d per week. I remember having to shovel snow off the roof during the bad winter of early 1947 because of fears that the roof would collapse. I nearly burnt the garage down once. I was changing a petrol pump on Major Lindsell's Morris Oxford and petrol spilled out over the lead lamp and caught fire. We tried all things to try and put it out. Someone sprayed water on it but all this did was to spread the fire. We tried sand from a bucket but it came out in one lump. Eventually, the fire brigade were called and a fireman simply licked his fingers and put it out by covering the end of the pipe to stop the flow of petrol. I had to have treatment for some burns. I can remember many of my colleagues during my time there. Arthur Bottoms was the foreman when I arrived. He owned the Triangle Café in Bridge Street which was run by his wife. Charles Cushing took over from him as foreman. I remember Roy Sanders. He lived at a house called Layston in Wymondley Road. Lance Sanders, who also lived in Wymondley Road, used to pick the apples from his trees, take them to the garage and sell them, yes sell them, to the staff. Bill Humphrey, a lovely man, was also there. I was told that he used to take Sir Ralph Delmé-Radcliffe to the railway station each morning in his Rolls-Royce and then bring it to the garage to clean it and park it until Delmé-Radcliffe returned in the evening. During the war Delmé-Radcliffe traded the Rolls in for an Armstrong Siddeley. As Bill lived next door to the garage he was always on call to attend breakdowns. He used the old Rolls with a crane fitted by removing the lid of the dickey seat. I've still got the box from the Rolls which was attached to the running board. His son, Donald, also worked at Sanders. He left to work at the Post Office in Walsworth Road for a while but then got a job at Stevenage Motor Company. David Humphrey, Bill's other son, was brought in by Major Clarkson to run the garage around 1959. David had been a rally driver during* the 1950s. Henry Barker was in the stores and Bill Blows was the blacksmith. Geoff Clark, who was a young mechanic, also worked there and married Bill's daughter, Jane. During the war Bill Blows was a prisoner in a Japanese concentration camp – all he was left with was his loin cloth. There was also an apprentice named Freddy Coutts. A chap called Eddie Calvert (not him with the trumpet) joined us for a while but soon returned to Coventry. Willy Woods, or Ginger, started at Stevenage Motor Company but, when the two companies merged, he was asked if he would transfer to Hitchin. We became good friends. Mike Anderson was General Manager in Hitchin in my early years, later replaced by Peter Badcock. David Louch, who worked at Stevenage Motor Company, had started as a fitter and had worked himself up to a management job. I also helped with the management but there was too much paperwork. The paint shop was upstairs and the only access for vehicles was by a lift which you raised by turning a big wheel. Later Willmott's replaced it with a ramp. [Robert Sanders remembers that the paint shop was originally downstairs but had to be moved upstairs when the ramp was installed.] The painter was Joe Froy. Kershaw's coach was parked next to the paint shop and I often sat in it to have my lunch. George Sharp, the banner maker, also had his studio upstairs. There were two generators downstairs as a backup to supply electricity for the garage. The new workshops were built over the road to service commercial vehicles – we had a contract to maintain ICL's fleet – but, soon after,

The generators used by Sanders to supplement the supply of electricity to the Walsworth Road garage.

all the repairs were carried out over there as the old workshops were too cramped. They also opened a petrol station in front of Desmond House run by Jack Tainch. He lived in the bottom flat of Desmond House. Ted Challis, who worked in the stores, lived in the top flat. Next to Desmond House were the lubrication and wash bays. We were told that one of the bays was originally used as the billiard room for the house. The mess room was in the basement. Ted Izzard had his paint spraying business in garages next door. I left Sanders when it was still in business so don't know how or when it closed.

I remember that BEWAC and Inchcape were involved towards the end. I then worked at Thomson's Garage in Queen Street for another 17 years until I retired."

William Woods

Here are Willie's reminiscences, known when he worked at Sanders as 'Ginger'. *"I was born in 1942 and, on leaving school aged 16, started a 6 year apprenticeship at the Stevenage Motor Company. Around 1963, when I was 21, I was asked if I would transfer to Sanders in Hitchin in exchange for someone that they wanted to train as a manager in Stevenage. I was happy to move and wasn't interested in management anyway. The job was alright at first - I started doing pre-delivery inspections on Morris Minors - but it was less enjoyable when they brought in the bonus scheme. I later worked on quality control and you wouldn't believe the dodges taken by some of the mechanics to get the jobs done quickly to earn their bonus. I remember one incident when a colleague was working on a big Mark 10 Jaguar. He adjusted the carburettor in the old workshop whilst the engine was running and the car gently moved forward, through the brick arch and knocked the sliding doors flat on the ground. The car was stopped before it reached the road and no damage was done to it at all. On another occasion a Morris 1800 rolled off the ramp, cracked the sump and knocked over a petrol pump. The hoops [chocks] hadn't been fitted to the ramp to allow the car to be moved manually*

whilst adjustments were made to the engine but the car was too close to the end and fell off. When I was doing a pre-delivery check I put my cigarette on the fabric covering protecting the bumper and it caught fire just as Major Clarkson arrived. I tried three fire extinguishers to put the fire out but none of them worked. I got a telling off from the Major. I remember watching a colleague fit a differential back-to-front on a Series 'E' Morris which resulted in one forward gear and four reverse! Major Clarkson, not a popular man, occasionally

William Woods, known as 'Ginger', servicing a Jaguar Mk2, in the 1960s.
(William Woods)

visited from Stevenage and kept trying to get a time clock installed. David Humphrey, the manager, resisted for several years but finally had to give in. I liked David who was very fair with the men. He had a bachelor pad on the island of Majorca and he once asked me to fit a new clutch to the Land Rover he kept there. I agreed so long as David would take the clutch, it was too heavy for me to take on the plane with my luggage, and that my wife and two year old child could join me – David readily agreed. We had a good week in David's flat but he never showed up with the clutch! Sanders always had a good reputation for after sales service. Whether it was new or second hand, if a car had a genuine fault, they would correct it without question free of charge. I left in the 1960s to join Shelvoke & Drewry – Colin Orchard and Alan Burling joined them before me – and I also worked at Zenith Motors, the Ford dealer in Stevenage."

Jenny Feaver (formerly 'Miss Potter')

Jenny, then Miss Potter, worked in the office from 1956 to 1962, lastly as secretary to David Humphrey. Here are Jenny's reminiscences. *"Mr Bernard took control of the business after Frank Sanders' death in 1949. He occupied the ground floor office behind the showroom. His secretary was Audrey Dilley (née Sell) who also prepared the job cards. The telephone switchboard was operated by Isobel Ashpool. Donald Humphrey, son of workshop manager Bill Humphrey, also worked in the office dealing with the accounts. I*

The interior of the Walsworth Road showroom with the large clock and "Miss Potter" at her desk in the office with the glazed door.

remember the visits by Major Clarkson who usually ordered me to 'bring a dish of water for Michael of Townsend', the way that he always referred to his pet dog. Major Clarkson also 'encouraged' me to leave my job when I became pregnant."

Desmond House and the New Workshop

Desmond House, originally known as Alpha Villa at 88 Walsworth Road, was built in the mid-1850s for Thomas Hall, a farmer; there is still a stone tablet, dated 1855, bearing his name, inserted in the boundary wall with the adjacent property known as Rose Cottage (where Stuart Sanders and his partners carried out their accountancy practice). Around 1888 it was bought by the Lines family. The '1913 Project' records that, in 1915, the property included *'a brick-built stable for two horses and a coachhouse'*. In 1923, the Lines family sold it to George Clark Payne, a 'gentleman of the turf', who used the proceeds of the winnings on a horse named *'Desmond'* to purchase the property, resulting in him changing the name to Desmond House. He also had a holiday home at Jaywick on the east coast but, tragically, was drowned in a boating accident in the 1930s. Frank Sanders bought the property from the Payne family, who had moved to nearby Verulam Road, in 1935, and it was transferred into the company's ownership in 1937 as recorded in the property register. Initially, the house was occupied by a tenant and, in 1939, some of the garden land at the rear was sold to the neighbouring builders, John Willmott & Sons Ltd, to enable them to extend their yard. In the same year a row of six brick garages were built by A E H Theobald for *£300 complete'*. Next to them was the body shop run by the Izzard brothers. The following period is uncertain and I have been unable to trace any documentary evidence to confirm either owner or occupants. Robert Sanders believes that, on an unknown date, the house was sold. It is recorded to have been run subsequently as a nursing home for the elderly. I was told that Robert Sumner Curling, the owner of Maydencroft Farm in nearby Gosmore village, was living there in the late 1940s and died there in 1949.

However, by the late 1950s, when it became essential for the business to acquire more space to service and repair cars, particularly commercial vehicles, the house and remaining garden were re-acquired and adapted to suit the business. Robert believes that it was re-purchased in the mid to late 1950s with more land for John Willmott & Sons Ltd as a condition of the purchase. A new low wall with two open entrances at each end replaced the original wall and wide single gate, which had included ornate iron railings that were removed and melted down for the war effort. Immediately in front of the house petrol pumps were installed on an *'island'*, so that cars could fill up from both sides of the pumps, cars

Sanders' advertisement for Exide batteries in a 1950's edition of the Hitchin Official Guide. Sanders were the Exide agents for the district.

Two Sanders' advertisements (left) for the Jaguar Mk VII, fitted with automatic transmission, in a local paper from 1956. (Right) for the Rover 3 litre in a 1961 issue of Hertfordshire Countryside

entering and exiting the site via a slight slope leading up from the new entrances. A glazed kiosk stood between the pumps on the 'island' for the attendant, who would fill your car with fuel - before the days of self-service - and where the cash till was located. The companies that supplied the fuel were Regent and later Burmah Oil. To the left of the house was a building with two bays; the first, possibly converted from the former garage or coach house (or even a billiard room), had a pitched roof and the second was built later with a flat roof but with an identical façade. One was used as a lubrication bay, fitted with a four post lift, and the other as a wash bay; they both had roller shutter doors. The house was converted into two flats for staff members. The upper flat was occupied by a salesman who worked at Stevenage Motor Company and later by Ted Challis and his family. The lower flat was occupied by Jack Tainch, a widower with two sons. He ran the petrol station. The basement was converted for use as the mess room where the mechanics and other staff had their tea breaks and lunch. Later, a large workshop was built behind the house fully fitted with lifts and oil fillers supplied by hose pipes on rollers. It was opened in 1961/2 and was originally intended for the servicing of commercial vehicles. Sanders had service contracts with several local companies, ICL probably the largest. However, soon after, cars were also serviced in these new premises. The orchard remained and was a source of fruit and fun. Robert remembers learning to drive in an old Austin 7, given to him by his father, by driving it round and round the car park.

These early post war years were the heydays for the modern business but times continued to change and major decisions had to be made about the future of the company.

A Sanders' business card

The End of the Road

In this the final chapter I will tell you about the last years of the business, including what happened to the various buildings and to some of the personnel.

Shareholders

On 22nd June 1962, the Stevenage Motor Company gained full control of Sanders Garage on the purchase of the remaining 10,060 shares from several members of the family giving them a total of 15,000 shares. At that time Ralph's three surviving sons held around 1,300 shares each, his grandson Bernard 3,000 and his daughters the remaining shares. Although Bernard, and also David Humphrey, Major Clarkson and Frank Broadribb were re-allocated a nominal 10 shares in the old company, which was finally liquidated on 19th November 1982, the family business now effectively ceased to exist. By the mid-1960s the business, then known as the SMC Group, being both the Hitchin and Stevenage garages, had expanded physically and commercially as far as they were able and a decision needed to be made about its future direction. Some have suggested that further expansion was, in any event, financially impossible because dividend payments to shareholders had not allowed the necessary capital to accumulate. Major Clarkson, Chairman of the Group, is reputed to have said *'We are too big to be small and too small to be big'* and his decision was to sell the company to a larger concern. Who instigated the

Sanders' advertisement in the Evening Post–Echo of the 1st December 1977, showing a Morris Marina and an Austin Princess in the showroom and offering £30 worth of 'Xmas Fayre' if you bought a car by 22nd December

negotiations is not known but, in 1965, the SMC Group was sold to BEWAC (British Engineering West Africa Company), a company that had its origins in Nigeria. They already owned several other garages in the UK, including Lathams in Leicester and Hinckley, and others in St Ives (Cambs.), Coventry, Mansfield, Bournemouth and Berwick-upon-Tweed. It is understood that they imported British cars, particularly BMC cars, into Africa in CKD or knock-down kit form for assembly. The UK dealerships also supplied cars to visitors from Africa on a lease hire basis. A new showroom was built at Stevenage in 1967. In 1970, BEWAC themselves, or certainly their UK motor dealerships, were taken over by Inchcape. The company was named by its founder, James Lyle MacKay, Baron Inchcape of Strathnaver, after the treacherous rocks 11 miles out to sea off the east coast of Arbroath, Scotland upon which engineer Robert Stephenson built a lighthouse, in 1811, which still stands and saves lives today. Inchcape owned many garages nationally, including the Mann Egerton chain which already had a dealership in Hitchin, on the corner of Queen and Bridge Streets, a possible reason why they chose to close Sanders Garage in 1979. The Mann Egerton garage building is currently occupied by a Carpetright store at one end and a branch of ASDA at the other.

Bridge Street Properties

It is believed that, by the end of the thirties, Sanders had sold all their Bridge Street properties with the exception of the

The former Sanders' car showroom at 13 (now 26) Bridge Street, now the 'Bridge Street Bistro'.

The replacement for the Carriage Repository and shop at 15 and 16 (now 28 and 29) Bridge Street, with the former Royal Oak public house on the left. (Stephen Bradford-Best 2013)

bus garage, built by them behind number 13, which was sold in 1953. I will record the fate of the other buildings in Bridge Street for completeness. The 'new' building, at number 13 Bridge Street (now number 26), built by Sanders around 1915, was later occupied by several small businesses, including B L J Hart, Agar Motors Ltd. and the Codicote Press, until it became a restaurant, firstly Claude's Crêperie and later the Bridge Street Bistro, which continues today. The range of buildings that comprised numbers 14, 15 and 16 (now numbers 27, 28 & 29) were subsequently home to various businesses, including Reed, Townsend & Co Ltd., Five Star Car Hire, Crone's furniture shop and Allender's DIY store, until the 1960s when they were completely rebuilt, winning a Civic Award in 1969. They continue to provide accommodation for small businesses. The Maltings, on the opposite side of Bridge Street, home to Slater Batty & Co in the early years of the last century, were demolished and the land used for parking buses until John Ray, of The Hitchin Property Trust Ltd., built his offices on the site. The Sale brothers' garage, further up the street near the river and which fell down unexpectedly, was later the site of government offices known as Crown House, built in the austere and unpopular style of the 1960s, and which has also now been demolished to make way for a smart residential development of apartments known as Maltings Court, named after the Lucas Brewery which once stood on this site.

You can still see the distinctive oriel window and roofline of the original Walsworth Road garage, now occupied by the Kwik-Fit tyre, exhaust and MOT centre.

'Sanders' Place' – apartments built on the site of Sanders' 'new' workshops and petrol station in Walsworth Road, previously Desmond House. (Stephen Bradford-Best 2013)

Walsworth Road Properties

A plan inside the rear cover of this book shows the extent of the Sanders' premises when their business closed in 1979 (although the shaded area also encompasses Izzards Coachworks). Following closure the original garage became a Peugeot agency under new ownership. Later, boats were sold from the showroom. The ground floor then became a popular tyre and exhaust centre, owned by Kwik-Fit, which still continues today, although the part of the building closest to the station, which gave access to the ramp leading up to the first floor, was demolished in 2000 to make way for a residential development of apartments. During a recent visit to the ground floor of the surviving building, with former Sanders' employee Geoff Clark, many features of the time when it was used as the garage workshop were identified, including the position of the pits, lifts and the offices. We also explored the basement where paint had been stored. On the first floor, unoccupied for many years, we were amazed to find that the original large timber framed carriage workshop and the original landmark brick built car showroom, one of the first in the country, both built in the first decade of the 20th century, remain intact and unaltered. Perhaps English Heritage should show more interest in these important buildings which have survived for well over a century and not be put off by the modern cosmetic additions to part of the front elevation and some metal cladding elsewhere.

The site opposite the original buildings, including the 'new' workshop, became a Renault agency under the ownership of the Walsworth Motor Company. They had Desmond House and the lubrication and wash bays demolished, and the petrol pumps and kiosk removed, to enable a new large car showroom to be built with a display area in front for second hand cars. It remained a car dealership for several years but, in 2003, the site was acquired by Stephen Howard Homes and all the buildings on the site were demolished and removed to make way for a residential development of apartments known as Sanders Place.

Other residential properties in Walsworth Road remain in family ownership, including number 58, next door to the original showroom, purchased by Stuart Sanders from the Stevenage Motor Company so that Bill Humphrey's widow could continue to live there in her retirement. It is now owned by Stuart's youngest son, Jonathan Sanders.

Personnel

Bernard Sanders was the last member of the Sanders family to manage the business. By the time it was sold to the Stevenage Motor Company in 1962, he had, then aged 57, been replaced by David Humphrey as manager, although Bernard did stay with the company until he retired in 1972, aged 67. The change of his role must have had an impact upon him, bearing in mind that he had, by then, worked for the company for more than fifty years, the latter 13 years in a senior role. However, the company would have gained from the regular customers that he had during his long association with the garage. In his later years, even after his retirement, he enjoyed delivering new cars for the company. Bernard died in retirement in 1990, aged 86.

The business finally closed in 1979 (the Stevenage site in April 1980) and the remaining staff had to find other employment. David Humphrey had left Sanders before the business closed. He suffered misfortune in his later years and died in 2001, aged 70. Others who also left before closure were John Brown who, in 1975, joined Thomson's Garage in Queen Street as their workshop manager, as did Arthur Scott as their sales manager. Jenny Feaver, formerly 'Miss Potter', who had worked in the offices in the late 1950s and early 1960s, also joined Thomson's Garage

Geoff Clark runs his car repair business in the workshops in West Hill, Hitchin, built in the 1920s by Bert Wells. (Stephen Bradford-Best 2013)

at this time and became secretary to Neil Thomson. When Geoff Clark left Sanders he started a car maintenance and repair business in Wells Garage, West Hill, in a building built by Bert Wells and his father in the 1920s and little changed since that time. William Woods is enjoying his retirement. Many other former employees survive and have, in the past, met up occasionally for a reunion and to chat about old times at the garage.

The Sanders' company, and the buildings it occupied in Hitchin, typify the transition of this type of family business from the Victorian world of horse drawn carriages, followed by the introduction of early motor cars through to dealing with the mass produced vehicles of the mid to late twentieth century, although, unlike Sanders, many were unable to master the new technologies, and incorporate the new trends in transport, and fell by the wayside. The business was started by a 'big' man, Ralph Erskine Sanders, and closed with his business being swallowed up by a 'big' national company. From the time that Ralph first established his business in Royston in 1875 until the Hitchin business closed just over a century later, Sanders were always ready to innovate and expand, to keep up with the times and to provide the local community, and later a wider area, with a comprehensive transport service - from carriages to cars.

Commissions: horse drawn vehicles

Commissions - horse drawn vehicles

The following are five examples of a variety of horse drawn vehicles, and one cart, built by Sanders commissioned by its customers - both for personal transport or for commercial purposes. I believe most of the terminology is self explanatory, although I should explain that the advantage of a Collinge axle is that it is cranked in the centre to allow the body to be fitted lower to the ground between the wheels.

Example 1

Date: *January 1906*
Customer: *W W Rowe Esquire, 133 Seven Sisters Rd N. Description: To a new Miniature Victoria Phaeton as photograph submitted, body hung on a set of best steel elliptic springs and Collinge axle, fitted with light Warner wheels and rubber tyres, painted dark green, striped in lighter green, upholstered in best Elans's green coach cloth, fitted with hood of best enamel leather, lined with cloth to metal trimming, knee boot apron the same materials to metal, small detachable rumble for page boy, best wood wings and dash, complete with pole and bar for pair, brass pole chains with spring hooks, good lamps, brass beads and furniture, and rein clips. Additional cost of wood wheels, detachable driving cushion, enamel hood overlaid duck, enamel leather apron to*

SANDERS & SONS will be pleased to send a copy of this price list to any *bona-fide* enquirer, free of charge.

The best and latest designs in all types of fashionable carriages are built throughout in SANDERS & SONS' Manufactory at Hitchin.

(3413)

Sanders' Miniature "Victoria" Phaeton.

A charming little carriage, designed expressly for use with a pair of Shetland ponies or a single animal of about 12 hands. The photo-engraving is from a carriage large enough to carry two adults, with comfortable accommodation. The body is suspended on Cee springs behind and elliptic springs in front and is mounted upon india-rubber tyred wire wheels. The hood is of the best enamel leather and the whole of the trimming is carried out in a really high-class manner.

Price (including rubber tyres) 55 guineas.

(Unsolicited testimonial.)

As to a Miniature "Victoria" Phaeton and Harness.   SPALDING, LINCS.

I am writing to say I had an opportunity of trying the Phaeton on Monday and am very pleased indeed with it and find you have carried out the instructions in every way most faithfully. I could not find fault in any way whatever. The short time I had it out it was greatly admired. The harness, too, is very satisfactory. I trust later on we may again do business in having another in different style.

A drawing of the Sanders' Miniature "Victoria" Phaeton, similar to the surviving carriage owned by Amanda Goodwin. (See 'Surviving Carriages and Cars')

metal bead and new large dust proof carriage cover with tapes.
Cost: *£69.l6s.0d (paid by cheque 20th January).*

Example 2

Date: *20th January 1906*
Customer: *Messrs W C Windover, Turrill & Sons, 22 Long Acre WC* [prominent carriage builders].

(4037)

A Sanders' Special "Eclipse" Car, similar to the one supplied to Messrs W C Windover, Turrill & Sons, on display outside Alpha Villa and (right) the vehicle as shown in the company catalogue.

Description: *To a new full size round cornered governess car, hung on long side springs and scrolls and Collinge solid flapped cranked axle, fitted with Warner wheels and lancewood shafts, varnished throughout in natural wood and upholstered in best blue cloth, complete with brass rein rail, axle wrench, whip socket and lamps. Cost: £16.0s.0d as agreed, plus 12/- railway charges (paid by cheque 27th January 1906).*

Sanders' Special "Eclipse" Car.

An exceptionally stylish and fashionable car with body hung on long grasshopper springs and collinge patent cranked axles. Its shafts are of finest lancewood and play easily to the knee motion of the animal as they are connected in fulcrum bearings. The deep panels are relieved with mouldings and are padded and upholstered inside. The interior offers the advantages of ample seating accommodation for four adults and a perfect balance can be easily secured whether one, two, three or four passengers are riding.

The various sizes each embrace the most approved dimensions adopted by us after much careful thought, and for durability, lightness and attractive appearance the Special "Eclipse" car is considered by many users to be the most perfect obtainable.

Price, in the natural varnished wood and handsomely upholstered to choice,

In Pony size, is 28 guineas.   Cob size 30 guineas.   Horse size 32 guineas.
If painted to choice of colours 2 guineas extra.
Excellent lamps from 30/-.

All sizes stocked ready for finishing in any colour.   Best grades of rubber tyres fitted at most favourable prices.

Many new carriages were supplied to this coachbuilder during this period which later merged with the original company, Charles S Windover & Co.

Example 3

Date: *13th February 1906*
Customer: *First Garden City Ltd. Letchworth.*
Description: *To a new strong contractor's cart built with all English oak framework, strong elm plank sides, strong plank hind door to lift in and out, loose side boards, projecting front to carry ladders, scaffold poles etc., solid iron axle tree, good high wheels hooped with bevel tyres and cart strongly ironed throughout, varnished throughout in natural wood and name lettered on front side board.*

(3462)
SANDERS' LADY'S "STANMORE" DOGCART.
An admirable low-hung cart; quite easy of ingress and egress; mounted on wheels of good height and possessing an exceptionally good driving seat which can be moved backwards or forwards in any position. Its lancewood shafts lie in rubber-lined bearings in front and are attached to relief springs at their hind ends and play easily to the action of the horse. The hind door is hung on "dropped" hinges and admits ample leg room for the occupants at the back when opened.
Its price in the natural varnished wood with coach-cloth cushions is :—
Cob-size 30 guineas. Horse-size 32 guineas.

As to a "Stanmore" Dogcart and new harness.
TUBRID, CAHIR, IRELAND.
The trap appears a nice one and the harness Mrs. ——— is much pleased with.

The Sanders' Lady's "Stanmore" Dogcart, similar to the one supplied to F E Ward Esq.

Cost: *£ 21.10s.0d (paid by cheque 8th March 1906).*

Example 4

Date: *24th February 1906*
Customer: *Messrs. H Fordham & Co. Baldock.*
Description: *To a new superior built coal dray complete with shafts, slipper and chain, wheel chain and axle wrench.*
Cost: *£29.0s.0d – delivered at Baldock station.*

Example 5

Date: *May 1906*
Customer: *F E Ward Esquire, The Manor House, Newnham.*
Description: *To a new cob-size Stanmore Dogcart as viewed and approved, body hung on two long easy side and cross springs and Collinge patent axle, fitted with lancewood shafts in fulcrums and Warner wheels, painted dark green upper panels and black body, wheels and underworks painted primrose with black stripes, upholstered in superior green coach cloth as chosen, fitted complete with fibre mat, leather whip socket, adjustable footrest and brass furniture to shafts.*
Cost: *£31.10s.0d. (allowance for old cart 10 guineas. – balance paid by cheque 15th June 1906).*

Example 6

Date: *2nd November 1906*
Customer: *The Right Hon. The Earl of Lytton, Knebworth*
Description: *To a new miniature round cornered governess car, hung on double elliptic springs and Collinge cranked axle. Fitted with light hickory Warner wheels. Varnished throughout in the natural colours and upholstered in light fawn corduroy. Complete with rein rail, whip socket and axel wrench.*
Cost: *£10.10s.0d, plus 15/- to deliver same at the Manor House (paid by cheque 7th December 1906).*

I was given a copy of a photograph by Clare Fleck, Archivist at Knebworth House, of what I believe to be this Governess Car. It is shown standing outside Knebworth House. On the reverse is written *'on the pony - Anthony (Lord Knebworth), standing in a riding habit - myself* [Pamela, Countess of Lytton], *in the Cart Hermione* [later becoming Lady Cobbold], *with Nanny Butler, at the Donkey's head - 'Little Ethel.' The donkey's name was 'Batchelor's Button'.* Clare also kindly allowed us to quote from an unpublished memoir of Hermione, Lady Cobbold, entitled *'Tales of Long Ago.'* She writes *An alternative to nursery walks was the donkey cart and sometimes nanny would take us for a drive. If we drove into Stevenage to do some shopping we would be allowed to get out of the cart and run up and down the six hills* [a collection of Roman barrows situated alongside the old Great North Road]. *I remember one occasion when the donkey chose to lie down in the middle of the road and a passer-by called out 'Light a fire under his belly!' The donkey was called Batchelor's Button and was a great character, living to a great age and acquiring the ability to open all gates on the Estate and come and go as he chose.'*

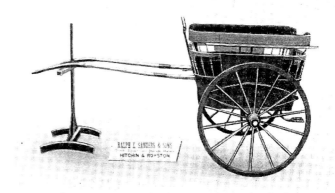

The catalogue entry for the Sanders' Miniature Governess Car as supplied to the Earl of Lytton.

A Sanders' Governess Car standing outside Knebworth House with Anthony, Lord Knebworth, on the pony; Pamela, Countess of Lytton, standing behind the pony; Hermione (later Lady Cobbold), seated in the cart with Nanny Butler; and 'Little Ethel' at the head of the donkey, 'Batchelor's Button'. (Knebworth House Archive)

Commissions: motor bodies

Commissions - motor bodies

The following are five examples of a variety of motor bodies built by Sanders commissioned by its customers for their personal transport or, in the case of the Adams Manufacturing Co., to be supplied to their customers. If recorded, I will first give the details about the motor body as recorded in the Day Book in the same way that I have for carriage customers.

Example 1

Date: *January 1907*
Customer: *S T Gresham Esquire, The Grove, Hitchin.* [Sam Gresham was a wealthy American].
Description: *To a new best quality limousine motor body as per quotation and specification of 30th November 1906, painted light green, striped in brilliant green and highly varnished, upholstered inside with hand buffed leather hides, silk laces etc., both seats and cushions spring seated, twelve lights all in selected polished plate 3/16" thick and bevelled edged, sliding ventilation at each fixed window, automatic spring winds at ditto, best polished brass furniture, lamp cantine (sic) set, Brussels carpet mat etc*
Cost as estimate: *£98.0s.0d (paid by cheque 25th January 1907).*

An Adams-Hewitt car, with a two-seater racing style body, probably built by Sanders. (Beds & Luton Archives Service. IG 8/4/1 p.184 (fig 298)

118

Example 2

Date: *19th September 1906*
Customer: *The Adams Manufacturing Co., Elstow Rd., Bedford.*
Description: *To a new racing type motor body as per particulars and dimensions supplied, painted and upholstered in standard green, with black mouldings, complete with bonnet, swing bolts, front and rear lamp brackets, hook and staple for supporting body when in a raised position.*
Cost: *£14.0s.0d.*

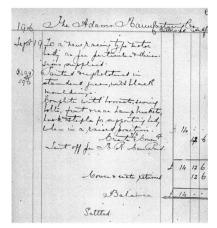

The Day Book entry, dated 19th September 1906, for the building of 'a new racing type motor body' for the Adams Manufacturing Co of Bedford.

Further similar racing bodies, and other bodies for motor vehicles, were supplied to this company during this period, including a new Victoria motor body *'as per blue print, painted dark green, striped in red'*.

The Adams Manufacturing Co Ltd was founded by American, Arthur Henry Adams, in 1905. Adams had married into the family of the founder of the Hewitt Motor Company of America and had originally planned to mass produce cars, known as the Adams-Hewitt (later simply Adams), in the UK using a Hewitt back axle. However, after a year, the company made all the components for the cars in their Elstow Road factory in Bedford, which was encircled by a perimeter test track. The first cars used a single cylinder engine mounted under the floor but these were soon followed by cars with engines mounted conventionally at the front; unconventional was the optional epicyclic gearbox

On checking the registration records held at Bedfordshire and Luton Archives, I could only find one Adams-Hewitt car with a 'Victoria' body. It was fitted to a 10 hp chassis and was registered with the number BM 603 to Richard Stephen Whiteway of Brownscombe, Shottermill in Surrey. Whiteway, born in 1845, was appointed to the Bengal Civil Service from 1868, serving in the North-Western Provinces. In 1893 he and his wife retired to England where he became a respected author on the subject of the Portuguese and their relationship with Abyssinia and India in the 16th century. Why did an Englishman who spent his career in India choose an Adams-Hewitt? Possibly because he knew that Adams exported their cars throughout the British Empire and that the British Government had presented one to the Emperor of Abyssinia.

It is difficult to be certain of the use made of the two Sanders built 'racing bodies', painted green, supplied to Adams on 22nd November 1906 but, again, checking the registration records held at Bedfordshire and Luton Archives, I found that two Adams-Hewitt 2-seater cars were registered shortly after supply, on 19th December 1906, with numbers BM 679 and BM 680, the only cars of this type registered within a six month period and never elsewhere two of them together. These cars are likely to have been fitted with the 10 hp 2 cylinder engine. The first was registered to John Clutton of Cranbrook, Kent and the second to Forster Delafield Arnold-Forster, later a Rear-Admiral who served in the First World War. Those familiar with the old car movement in the latter half of the 20th century will be familiar with the names Cecil Clutton and Nigel Arnold-Forster but whether their families ever owned an Adams-Hewitt car is not known. Bedfordshire and Luton Archives have kindly allowed us to reproduce a photograph from their collection of a 2-seater Adams-Hewitt car, possibly (probably) with a Sanders body.

I will continue this section with a more detailed story of three of the motor bodies supplied by Sanders based on information either from the Day Books or from other sources, together with facts about the customer, his family and/or his career.

Example 3

Dr William Briggs - Panhard Landaulet

Date: *18th January 1908*
Customer: *W Briggs Esquire, Owlbrigg, Chaucer Rd., Cambridge*
Description: To a new coach built dee-fronted landaulette body constructed of bone dry English oak framing, panelled in mahogany, to dimensions suitable for a 35 hp Mercedes chassis, fitted with folding head complete with automatic action, patent slam locks doors, long platform steps of good width having shaped covers to protect gears with opening doors and shut iron steps, accumulator case in front body protected with long planished steel wings, the space under inside seat reserved for tools with door at back, the roof extended over driver's seat and fitted in front with a patent Royal windscreen with plate glass windows, upper part adjustable to any angle and with outside brass supports carrying roof at front, luggage rail fitted on top and enclosed with a small mesh guard on all sides, strapped windows fitted in front, at doors and in quarters at rear, the body and chassis painted throughout in best style, dark coach green, picked out black,

A Sanders' landaulette body fitted to a Panhard chassis, similar to the one supplied to Dr William Briggs in 1908.

edged white and highly polished, the interior luxuriously upholstered in superior Paris cord finished with wide silk laces, wallets on either sides, doors fitted with tip up sprung seats fully trimmed to match, best cushion and squab sprung seated, elbow rests hanging from swivel joints, inside automatic curtains to all windows made up in green silk hung complete with all tassels, plain pile Brussels carpet to lower part of floor, the interior fittings of real horn, best silver plate metal parts and fittings, polished plate glasses with bevelled edges fitted in oak frames covered in Elans cord cloth, outside seat trimmed in hand buffed green motor hides both seats spring seated, the hood covered in selected cross grained enamel leather, platform brass plated with fluted angles and covered with footboards, gear steps etc. Pyramid India rubber, brass cornice round roof, shell pattern door handles etc. **Cost: £155.0s.0d.**

I found that Dr William Briggs was Principal of University Correspondence College in Cambridge and that he died on the 19th June 1932. The University Correspondence College was founded by Briggs in 1887 and pioneered the provision of support to those studying with the University of London at a distance. Briggs' college provided a correspondence tuition scheme by post, along with face to face day and evening teaching in London and Cambridge, short residential schools and the production and sale of specially written texts to help students. His very effective system is a common feature of the range of services used in modern distance education. At a more general level, between 1887 and 1931, 39,326 external students are recorded as passing University of London examinations, with some 10,000 gaining Bachelors or even Masters Degrees.

Example 4

Lord Carbery - Cottin et Desgouttes 40hp Sports Tourer

Around 1911 Lord Carbery commissioned an elegant sporting tourer on the 40hp chain drive racing chassis. Sadly, I do not have precise details of the body as no Day Books dated after 1909 survive. But who was Lord Carbery?

John Evans-Freke, 10th Baron Carbery of Castle Freke in the County of Cork (Irish Peerage), was born in 1892. He first married Jose Metcalfe in 1913 but they were divorced in 1919. Next he married Maia Ivy Anderson in 1922 but she was killed in a flying accident in Kenya in 1928. He then married June Weir Mosely in 1930, who outlived him by ten years. In addition to Cork, he had a home in Pebble Beach, California. In 1920 he renounced his title and, in 1921, legally changed his name to John Evans Carberry, sold Castle Freke and moved to live in 'Happy Valley' in Kenya. He was a skilled pilot, as early as 1912, and, during the First World War, he joined the Royal Naval Air Corps with his own plane. He also represented Britain for the Schneider Trophy air race. He became part of the "jet-set" and, it is alleged, squandered a lot of the family money. He died on Christmas Day in 1970.

Sanders purchased several chassis made by the French company Cottin et Desgouttes from their British agent, the Cambridge Automobile Company. In fact, it is believed that Sanders built the majority of the bodies in England for this imported chassis prior to the First World War. In a letter he wrote to C L Grace Esq. in 1943 Frank says '*well remembers designing a special type of sporting body for powerful fast Cottin Chassis. This Chassis was chain driven, had demountable rims and the change speed lever was on the outside of the body.*' He goes on to say '*... it was sold by the Cambridge firm to a customer at or near Bury St Edmunds ...*' and '*actually there were only one or two of the Cottin Speed Models imported in this Country, they were built in Lyon.*' In a letter to the same recipient in 1944 Frank says '*The special body work was supplied by our firm to the order of Lord Carbery, when he was at Cambridge, and the writer well remembers going down there and taking his Lordship's instructions re the coachwork.*' In a letter to motoring historian Anthony Heal (whose family owned the well known furniture store in Tottenham Court Road, London), in 1946, Frank repeats the information about taking Lord Carbery's instructions in Cambridge, apologises for not having the old time records as they were given up '*when the call was put out for the*

*The French Cottin et Desgouttes 40hp chain drive Sports Tourer supplied
by Sanders to Lord Carbery in 1911*

The Cottin et Desgouttes with Roy at the wheel and young Bernard beside him in the passenger seat

surrender of paper during the War years ...' and lastly says '... *we speak very highly of the performance of these French built chassis*'" It is believed that Heal owned a similar car in the 1920s, hence his enquiries to Sanders. The copies of the letters from Frank to C L Grace and Anthony Heal were kindly provided by Oliver Heal, Anthony Heal's son.

Example 5

Charles Poston – Napier Landaulette

Charles Poston lived at Highfield, Stevenage, Hertfordshire. The house was in the hamlet of Pin Green which was absorbed by Stevenage New Town in the 1950s. Part of the estate survives as Hampson Park, named after athlete Thomas Hampson who won a Gold Medal at the Los Angeles Olympics in 1932. Poston, a solicitor with a practice based in the High Street and a JP, was married to Clementine and died in 1914. His daughter Elizabeth, born in 1905, moved with her brother and mother to nearby Rook's Nest, childhood home of the well known novelist E M Forster and the inspiration for his novel Howard's End, where she stayed for the rest of her life. Elizabeth, who no doubt travelled in her father's Sanders-bodied Napier, became a highly regarded composer and musicologist and had a distinguished career in radio broadcasting being one of the team that founded the Third Programme (now Radio 3). There were early links between the Poston and Forster families and Elizabeth stayed friends with Forster, known as Morgan, until his death in 1970. Elizabeth died in 1987.

In 1911 Poston chose the 15 hp Napier chassis for his motor car made by D Napier and Son at their factory in Acton Vale, North West London. Napier had been building cars since 1900

The Sanders' landaulette body fitted to a Napier chassis supplied to Charles Poston in 1911, with Frank Sanders holding the door.

for a mile of almost 30 knots. They later became famous for their aircraft engines, especially the Napier Lion, the most powerful engine of its day. The retail cost of Poston's chassis was £414:15s:0d and he paid a deposit of £130:0s:0d. It is recorded that Sanders held a £60 credit allowance for an Arrol Johnston Tourer and Poston paid the balance due for the chassis shortly afterwards. A landaulette style body was chosen not dissimilar, but more elegant, to the taxi body used by W & G Du Cros of Acton on their 15 hp Napier taxicabs between 1909 and 1911, some 900 in service before the Great War. It took about two months to built and fit to the chassis. It is also recorded that the net cost of the chassis paid by Sanders was £352:16s:0d, a third paid when the chassis was ordered on 15th May and the balance of £235:4s:0d when the chassis was collected on 13th June 1911.

The specification of the body is detailed in a typed extract from the Day Book made by Frank in August 1942; sadly the actual Day Book has not survived. The details include '...*seating accommodation for 3 passengers on the back seat inside and each upon the two occasional seats; outside accommodation for one passenger beside the driver.*' It goes on '*The hood fitted with patent spring action and opening from the front of door. Large window in the front to drop, also those in the doors; folding window-carriers provided on both sides, all frames made from figured walnut and polished, glazed with Polished Plate. Fixed extension over driver's seat, top arranged*

and had enjoyed great success in motor racing with their larger engine cars, including winning the Gordon Bennett Cup in 1902, their car being painted in the colour which became known as British Racing Green. They also built marine engines and, in 1905, their boat Napier II set the world water speed record *to carry light luggage and surrounded with strong luggage rail with wire mesh guard. Windscreen fitted in front, the top part to swing open at any angle. Chassis completed with the necessary step stays and running boards covered with aluminium and finished off with brass fluted surrounds; wing stays and best steel wings, those in front with outside valences; suitable brackets for head lamps. Side steps to front seat; high wind doors at the sides.'*

When it came to colour choice it says *'The car was painted in dark Green, usual parts Black, relieved with Red Stripes and highly varnished.'* Then for the passenger compartment it says *'Upholstery in material chosen and interior fitted with swing arm rests, drab silk spring curtains, speaking tube with long flex and brass trumpet, Brussels pile carpet on floor, cigar and card trays; main cushion spring seated; polished wood filets along tops of doors and around front windows, back light in hood, best nickel and white ivory inside fittings.'* Finally it says *'Hood covered with specially selected enamel hide; chauffeur's seat covered in Green Motor Leather and the front floor boards with aluminium plating; luggage straps for roof rail, lettering number plates, etc. complete £175:0s:0d.'* There were also some extras comprising *'Hall's Patent Window Flap fitted £1:15s:0d – a new set of best linen trimming covers with fasteners fitted £2:18s:0d – a strong luggage carrier to fold fitted at the rear; straps for fastening £5:5s:0d – a new proofed cover for hood to protect trimming from dust £1:10s:6d – supplying and installing electric light*

fittings and battery complete £3:15s:0d. This makes a total of the body of £190:3s:6d.' Poston paid £150:0s:0d on account by cheque on 14th August and the balance of £40:3s:6d in settlement on 21st August. The cost of the body and the chassis totals £604:18s:6d (or approximately £55,000.00 in 2013 values). I have no idea what Poston thought of his new car, or how far he travelled in it, but sadly he could not have enjoyed it for long because, as mentioned above, he died three years later.

A freehand drawing by William Sanders (then aged 13 years), made in 2012, of a carriage wheel. He is the great, great, grandson of Ralph E Sanders.

Surviving Carriages and Cars

The title of this section was chosen early on when writing the book and implies that I will tell you about a variety of both surviving carriages and cars but, despite strenuous efforts, I have been unable to trace any surviving motor vehicles fitted with a Sanders' body.

It has been suggested to me that should the motive power for your carriage wear out you simply put another animal between the shafts but if your internal combustion engine, and related mechanical parts, wear out you throw the whole car away and buy another - perhaps an extreme course of action but a plausible explanation for my lack of success. There is little doubt that, with the great advances in motor technology over the decades, owners did discard their old cars and replace them with the latest model. Tracing cars has the added problem of museum and car club records being maintained primarily by using the names of the chassis manufacturer rather than that of the coachbuilder whose name may or may not be recorded at all. I am still hoping that, following publication, an owner of a Sanders' bodied motor car will come forward and probably say "I have one: why didn't you ask me?"

However, I am very pleased to say that we have traced eleven surviving Sanders' horse drawn carriages of a variety of styles, all cherished and used by their owners. My search took off after I was contacted by Society member, Bob Prebble, who attended the Worstead Festival, Norfolk in 2010, with a friend, Audrey Griffiths. It was there that they met Sanders' carriage owner, Amanda Goodwin, who Audrey had first met at the previous year's event. Amanda gave us many contacts in the carriage driving world. They included well known carriage driver, Sybil Atkinson, who gave us many more contacts. In addition, replies to a successful appeal in *Carriage Driving* magazine put me in touch with further Sanders' carriage owners. One who contacted me was Vivien Hampton who lives only 12 miles from Hitchin and who gave some very useful advice. I have included information about those surviving carriages on the following pages, the brief descriptions written by the owners themselves who have also provided the photographs. A further Sanders' carriage is believed to be in Canada but, to date, we have been unable to contact the owner. Also, there are members of the Sanders' family who claim that there is a Sanders' carriage in the Royal collection, possibly a Dog Cart at Buckingham Palace, but, so far, it has not been located.

I will also record again my thanks to all those people with whom I have had contact in the world of horse drawn and motor driven vehicles for their enthusiastic support for my project - everyone went out of their way to help.

Governess Cart - David Rooney, Curator of Transport, Science Museum, London

"As a practical and convenient personal vehicle, this type of small pony trap was used extensively in the countryside. Light and easy to maintain and control, pony traps were in general use from late in the eighteenth century to early in the twentieth century. This typical example was donated to the Science Museum in 1938 following a letter published by one of its officers in The Times. In it, the museum asked for contributions to its horse-drawn vehicle collection, using the publicity the newspaper could generate *'to encourage public interest and help in this subject'*. A Mrs A.M. Taylor was quick to write offering her governess cart, telling the museum that *'it was in use when my children were small'* and describing it as *'a nice one of its kind'*. The museum was delighted to accept the cart. Mrs Taylor consigned it by the Great Western Railway on 26 August 1938, and it has been part of the museum's road transport collection ever since."

(Top left) The Governess Cart in the Science Museum's store at Wroughton, Wiltshire.
(Bottom left) The upholstered seat of the Governess Cart.
(Right) Sanders' chassis plate fitted to the Governess Cart.

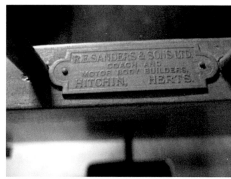

Governess Car - Paula Martin, House & Collection Manager, National Trust Carriage Collection, Arlington Court, Devon

(Above) The Sanders' Governess Car in the National Trust Collection. (Below) The rear of the Governess Car.

"Our 'Empress' Governess Car was built by Sanders' around 1900. It was given to the National Trust Carriage Museum in 2004 by Mrs S. Swannack who had used it for driving competitions up until the 1960s. It had been in her family for a number of years and was possibly bought by them directly from Sanders'. Governess cars, sometimes called tub carts for obvious reasons, have a deep tub-shaped body with access through a door at the back, and seats each side, to carry four. They were often used by a governess (hence the name), or possibly a mother, to take the children out for a drive. She would drive sitting diagonally in the back right-hand corner which was not particularly comfortable. The cranked axle allows the body to be mounted lower, dropping the centre of gravity and making it difficult to overturn. It has rubber shod wheels and is on elliptic springs. This design is known as a spindle back car because of the turned wood supports around the body. The body frame is made of oak, birch and walnut."

(Above) A side view of the Governess Car. (Below) Believed to be Mrs Swannack and her Shetland pony 'Midnight of Woodbury' driving the Governess car in competition.

Governess Car - Doreen Tindale, Abbeyfield Shetland Pony Stud, Rugeley, Staffordshire

(Above) Doreen Tindale in her Sanders' Governess Car. (Below) The Governess Car in front of Weston Park, Staffordshire.

"I believe that I am the third owner of my Governess Car. We are the 'Abbeyfield Shetland Pony Stud' established in 1978. Both my husband and I are international judges. In 2005 I was asked to judge the Shetland Pony Stud-Book Society's breed show in Aberdeen where we spent our time with Kay Gibb. She knows that I drive Shetland ponies as she also had as a child. Kay took us to an old lorry where inside there stood an old Governess Car which her father had purchased some 45 years before from an auction. He was told that the Governess Car was built for a rich land owner on Orkney around 1907, possibly the Sinclaire family. Kay's father then became the second owner. Weeks later Kay phoned, she was travelling south and asked if I would like the Governess Car. I replied "Yes, please!" Now we are the third owners. We made a trip to Wellington Carriages for a new floor and service and off we go – carriage parades, pleasure drives, etc."

(Above) The Governess Car in an inside riding school. (Below) Doreen Tindale in her Governess Car with three happy children

Spindle Back Gig - Malcolm Salter on behalf of Debbie Charlesworth, Holmes Chapel, Cheshire

"The spindle back gig by Sanders of Hitchin was purchased for Mrs Debbie Charlesworth by her husband as a 50th birthday gift in 2009, after she had seen it for sale in the Reading Carriage Sales. It was unsold and it was later bought from the dealer vendor privately. The Gig had been recently completely renovated in yellow and black by Fairbourne Carriages of Kent. Unfortunately, prior to that there is no information as to its whereabouts. It is a beautifully balanced cut under spindle gig with a very elegant outline and is suitable to be driven with a 12.2hh to 13hh pony. The gig is currently being exhibited very successfully in private driving and Concours D'Elegance classes with a Welsh Sec B pony by Mrs Nicola Salter for Mrs Charlesworth. The combination is a regular winner on the county show circuit, securing a highly contested Horse of the Year Show qualification for 2013."

(Top right) Nicola Salter driving 'Brynllwyn Quick Flash' to the Sanders' Spindle Back Gig
(Bottom right) Nicola Salter driving 'Brynllwyn Quick Flash' to the Sanders' Spindle Back, cut under, Gig winning the Private Driving Class at Lincolnshire County Show.

Miniature Victoria Phaeton – Amanda Goodwin, East Dereham, Norfolk

(Above) The Sanders' Miniature Victoria Phaeton owned by Amanda Goodwin. (Below)The carriage at the Royal Norfolk Show in 2008.

"The Brighton Miniature Victoria Phaeton four-wheeled carriage is believed to be still in its original condition and was built by Sanders in 1890. The carriage is lightweight and beautifully proportioned with its fine wire spoked wheels that are black and cream lined. The body of the carriage is also black and cream with gold coach stripes and brass fittings. The dashboard is of black crocodile effect leather with its curved and sweeping design. It has its original black leather hood with a brass edged bevelled glass window in the rear. Internally the carriage is spacious and lined in soft tan felt velvet with tan leather button backed seats. The carriage has oval shaped lamps attached in brass holders at either side of the dashboard. The shafts for this carriage were made at the same time as the carriage and are called swan necked; these are very much more elegant in design than straight shafts. They also have matching brass fittings with black patent leather to match the carriage. The carriage is still used by us for weddings etc. We also take it to shows, including many British Driving Society events where it is shown in the vintage carriage classes with our pony, *'Silver Star'*. Our little carriage has been pictured, and had many write-ups published, in Carriage Driving and British Driving Society magazines and in many wedding magazines."

(Above) The carriage drawn by our grey Welsh pony, 'Silver Star', with a removable seat in place for the driver as used for weddings. (Below) The Miniature Victoria Phaeton with the hood erect.

Governess Car - David Biles MBE, Northwood, Isle of Wight

"My family have farmed on the Isle of Wight for many years. The governess car really belongs to my son, Sam, who inherited it over 30 years ago. It was previously owned by a local butcher, Sammy Moul, who was well known in the area. On a Saturday night he would head home from the pub in his pony and trap rather the worse for wear with the pony in charge, indeed the pony would often tip him out on the way. I often chatted to Sammy who was convinced that we had named our son after him; so much so that he left the cart to Sam when he died. To be honest, we didn't have Sammy in mind when we named our son. The Governess Car is in good condition and has been used at many local events over the years, including the *'concours d'elegance'* at the Queen's Diamond Jubilee celebrations here on the Isle of Wight in 2012."

David Biles' Sanders' Governess Car at the Isle of Wight Show in 2012.

David Biles with his Sanders' Governess Car

Varnished Well Bottom Gig - Anthea Birch, Crickhowell, Powys

"We have owned the gig for about six years. The person we bought it from had only owned it for six months and then found that they had no use for it. We had it renovated, taking a year, and have used it since mostly driven by a young friend, Rhian Ralph. The gig is always admired wherever we go."

(Below) The Varnished Well Bottom Gig owned by Anthea Birch and driven by a young friend, Rhian Ralph.

Four Wheeled Dog Cart - Vivien Hampton, Benington, Hertfordshire

"In 2001 I was looking for a new traditional vehicle for my 42" Shetland pony and was told of the Dogcart at a farm in Bedfordshire. I went with my friend Sally Dashwood-Evans, who grooms for me, to see the vehicle, built by Sanders around 1910, and fell in love with it straight away. Although in a sorry state, all the wood was sound, I knew that Barry Wheelwright would be able to bring it back to its former glory, so it was purchased and taken to Barry for restoration. We have successfully shown in Private Drive and Vintage Vehicle classes since then. In 2002 we were invited to take part in 'All the Queen's Horses', at Windsor, in celebration of the Queen's Golden Jubilee and in 2006 we went to Sweden for the International Shetland Show where the turnout was awarded Driven Champion."

(Above) Vivien Hampton's daughter, Abigail, is seen here driving 'Makalan Mac' to the Dog Cart at Smiths Lawn 2006. (Left) The Sanders' hub cap.

Varnished Well Bottom Gig — Anna Mann, Upwell, Norfolk

"Ever since I began to drive a Shetland pair, I had wanted to put them in a Cape cart, a two-wheeled carriage with a pole, so as to be drawn by a pair or, indeed, by a team. It soon came about when I purchased a 'Bennington Baby Buggy' for 11-12 hh with optional pole for driving a smaller pair in Cape harness. But that was not the end of it. The next thing was to adapt an original vehicle. We found the well-bottom gig by Sanders and Sons twelve years ago on a farm in Suffolk. It had not been used for some time and exactly fitted the bill; the shafts ran outside the body, so they were easily taken off, and there was a cranked axle which gave a little extra room for the pole. Robin is good at woodwork and he reshaped a larger pole to suit. He was lucky to find, at Reading Carriage Auctions, a very pretty American buggy yoke with brass fittings. For the ironwork we went to Philip Holder, at Wellington Carriages, who made the pole housing and also put on the hooks that the swingle-trees hang from. The result is another light and comfortable carriage. We have shown it a couple more times, always coming top of the line, but the ponies have now retired."

(Left) Anna Mann just coming out of the ring in her Sanders' gig at the Royal Norfolk Show in 2011, having won the Open Pairs and Tandem class. (Right) The rear of Anna Mann's gig showing the cranked axle.

Spindle Back Gig — Denis & Nicola Fisher, Charente, France

"My wife Nicky and I acquired the carriage, a Ralph E Sanders & Sons Ltd of Hitchin Spindle Back Gig, built circa 1898, with 40 inch dished 16 spoke wheels, built as a pony carriage and not adulterated in any way, via Show Producer, Gary Docking, in 1990. It was put to our Welsh Section B Pony named Granby Rowan, stable name 'Ashby'. The turnout was driven by Nicky and it quickly became a prolific winner on the Private Driving show circuit, including numerous Thimbleby & Shorland Qualifiers for the British Driving Society (BDS) National Driving Championships; Show Champion at Hickstead; numerous firsts at the South of England Showground, Ardingly; the British Driving Society Show at Smiths Lawn; repeat performances at Hickstead, Devon County and Surrey BDS Shows; and numerous wins at Area Shows around the country. Sadly, 'Ashby' has now passed away but we still have the gig with us in France and hope to use it again soon."

Nicola Fisher, driving with 'Ashby' to the Sanders' Spindle Back Gig, when she came first at the South of England Showground, Ardingly, West Sussex.

Four Wheeled Dog Cart — Richard Lanni, Bridge of Earn, Scotland

"I acquired the Dog Cart some 20 years ago and was made aware of its whereabouts and availability through Barry Wheelwright, Hollywood, near Birmingham, on a flying visit when my wife and I were on our way to Normandy for a holiday in France. I was desperate to find a suitable vehicle for our two Grey Shetland ponies who were going really well as a pair and so we hot footed it to view the Dog Cart. It was exactly what I wanted and in the words of the then owner "could be pulled by a pair of Labradors!" So much to my wife's indignation, I bought it on the spot and it's fair to say we had a very 'budget' French holiday as a result. The Dog Cart was painted Black with Gold Lining and in good condition, however Barry arranged to have it uplifted and we changed the colour scheme to Oxford Blue with Cambridge Blue Lining which set the Grey ponies off beautifully. The Carriage has recorded many successes in the show ring, notably at the Highland Show where we have won the Shetland class on too many occasions to mention, firstly with the pair of Greys and then subsequently with my pair of Black ponies, all up to standard height - 40 to 42 inches. Almost inevitably this led to Team Driving and Barry Wheelwright fashioned a Team pole, with the neatest little crab end and leader bars to suit, and the vehicle particularly suited that turnout with a good seating position to view the leaders. The Dog Cart has given us much pleasure as a family and is beautifully constructed in direct proportion to the small ponies, a particular art that Sanders had honed to perfection.

(Left) A rear view of the Sanders' hub cap and (right)showing the whip socket..

Side and rear views of the Four Wheeled Dog Cart owned by Richard Lanni

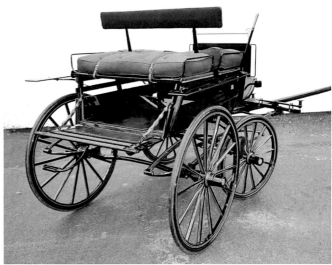

Kershaw's Coach
by Derek Wheeler MBE

For a good number of years a local sight was available to the observant in the first floor windows of Sanders' Walsworth Road showroom in Hitchin. From the 1951 Pageant until the early 1960s, and it is impossible to be specific with dates, what was known as Kershaw's coach was quite visible. One became used to seeing this wonderful example of the coachbuilder's art in such an appropriate location, the premises of one of Hertfordshire's foremost coachbuilders.

This canary-and-black road coach, lettered in Kershaw's style, often driven by George Mossman, had been one of the stars of the Pageant and had been seen much over the decade at local shows but was it the genuine article and where did it go?

In the striking period oil painting by Shayer in Hitchin Museum, the coach is not yellow. Generally mail coaches were dark red below the waist with a similar colour for the wheels. Was this the original coach, painted by Sanders' craftsmen especially for the Pageant after a long period in hibernation or was it another coach brought in and flying under false colours?

When Kershaw gave up his business and sold his 36 horses in 1850, what happened to the coach? Did Odell's ultimately purchase it and hire it out for fun trips to the Derby or Henley Regatta? Was it purchased by a local worthy and kept for sport and used by a member of the Four-in-Hand Club? There are period programmes in Hitchin museum for early 20th century coronation celebrations and a coach appears in these programmes, driven by Mr Gosling. A coach can be seen in the Market Place in some of Latchmore's photographs of the same coronation celebrations. One can also be seen in the Blake Brothers' black and white film of the same event. Was this a coach known as a private drag owned by the Goslings, bankers of Wellbury House, or was it Kershaw's coach emerging from half-a-century of mothballs? Is there anybody connected with Sanders who can answer these questions?

What happened to it is also a mystery. John Scorer, of Hitchin Historical Society, tried to follow the trail after it emerged that it was purchased by Collins Antiques in Wheathampstead. Various rumours were checked out. Some said it had gone to a brewery in Wales for publicity purposes. A story which was told to a former Assistant Curator at Hitchin Museum back in the 1990s was that someone had seen it on a derelict film set on an unspecified island in the Mediterranean. Brian Sewell, the art historian and journalist, visited a carriage museum as part of a televised art tour in Italy. A yellow and black English travelling coach was featured. David Hodges, Curator of Hitchin Museum, tried to follow up this sighting and met with a wall of ignorance as to its origin. Does anyone know the truth? Even present day carriage and coaching enthusiasts have drawn a blank on this enquiry. Wouldn't it be wonderful to have this bit of old Hitchin on display in our new Museum!

Kershaw's Coach standing outside Sanders' Garage in 1951 when it took part in the Hitchin Pageant. Clough Park, the Offley farmer, is on the front left-hand horse with Geoff Parrish, a farmer from Willian, on his right. The coachman, standing by the horses, is George Mossman of Luton. (Clough Park)

SOURCES

The main source of information and photographs for this book is the Sanders Family Archive which includes some Day Books (1906-1910 era); a Register of [Property] Documents; a Register of [Company] Members; a Timber Register; a small collection of Legal Documents (relating to apprenticeships and property dealings); catalogues for the ranges of carriages and motor car bodies built and sold by Sanders; Testimonials from customers; some advertisements and 'flyers' relating to services offered or items sold by the business; and a photographic collection, mostly black and white, which includes the buildings used by the business, the vehicles built and sold, and of family members. Most of the information in the book has been derived from this source, augmented by those written and internet sources listed below and all the photographs have been derived from this source with the exception of those that are attributed to others recorded against the photograph's captions. I have also quoted, or included information from, newspaper cuttings from the archive but rarely are the names of the newspapers, or specific dates, known.

Printed Sources – books

Douglas, P. and Humphries, P.: *Discovering Hitchin* – Egon Publishers Ltd.,1995

Field, R.: *Hitchin A Pictorial History* – Phillimore & Co Ltd., *1991*

Fleck, A. and Poole, H.: *Old Hitchin* – Phillimore & Co Ltd. 1999

Foster, A.M.: *Market Town: Hitchin in the nineteenth century* – Hitchin Historical Society, 1987

Gadd, P.: – *Fifty Years of Change in Hitchin* – AuthorGraphics Ltd.,1980

Georgano, G.N.: *The Complete Encyclopedia of Motorcars 1885 to the Present* – Ebury Press, 1973

Hamdorf, D.: *Seventeen Taxis'* – Ellison's Editions, 1983

Hine, R. L.: *Confessions of an Un-common Attorney* – J. M. Dent & Sons Ltd.,1945

Lane, A.: *Austerity Motoring 1939-1950* – Shire Publications Ltd.,2010

May, T.: *Victorian and Edwardian Horse Cabs* – Shire Publications Ltd., 2010

Morrison, K.A. and Minnis, J.: *Carscapes The Motor Car, Architecture and Landscape in England* – Yale University Press, 2012

Ralls, S. & J.: *Trouble with Traffic* – Royston & District Local History Society, 2007

Sillance, F.: *20th Century Royston. Volume 1 - the First 50 Years'* – Royston & District Local History Society, 1993

Taplin, V. and Stewart, A.: *Two Minutes to the Station* – Hitchin Historical Society, 2010

Whitmore, R.: *The Ghosts of Reginald Hine* – Mattingley Press, 2007

Wildman, R. and Crawley, A.: *Bedford's Motoring Heritage* – The History Press, 2010

Printed sources – Magazine Articles

The Coach Builders', Wheelwrights' & Motor Car Manufacturers' Art Journal: *June 1905*

The Motor: *21st September 1915*

The Autocar: *20th December 1919*

The Commercial Motor: *11th November 1930*

Herts. & Cambs. Reporter: *1st September 1933* – Hine, R. L. – Obituary of Ralph Erskine Sanders

Garage and Motor Agent: *9th October 1943*

Hertfordshire Countryside: *Autumn 1960*

Hitchin Historical Society

David Humphrey's memories, recorded in 1997

Hitchin Museum

Loftus Barham and Lawson Thompson scrapbooks
Map collection
Transport Box (ref.8207/166)
Road Transport Box (ref. 9446 (2-22) Sanders Garage

Hitchin Library

Printed copies of census records
Hitchin Directories
Hitchin Town Guides

Hertfordshire Archives and Local Studies

Vehicle registration records
Hitchin Urban District Council rate books
Kelly's Directories
Herts. & Essex Trade Directories
Hitchin Tithe map and award 1841-1844

Bedfordshire and Luton Archives and Records Service

Vehicle registration records
Photograph of Adams-Hewitt car

Knebworth House Archive

Correspondence between William Wilson, estate manager, and Ralph Sanders around 1920
Photograph of family with Governess Car

London Transport Museum

Photograph and information about the Rational cab

The National Archives

1910 Valuation Act field books for Hitchin (available on-line at www.Hitchin1913.org.uk..)

National Trust Carriage Collection, Devon

Details and photographs of Sanders' Governess Car in their collection

Science Museum

Details and photographs of Sanders' Governess Car in their collection

ZLS London Zoo

Photographic collection

Internet sources

www.ancestry.co.uk
www.thebunt.co.uk
http://www.hitchin1913.org.uk
www.nonsequitur.freeforums.org
http://www.ltmuseum.co.uk
www.ampyx.org.uk/lcountry/garages/garage_hn.html

Also Wikipedia, with corroboration from other internet sources.

Hitchin Historical Society

The Society aims to increase and spread knowledge of the history of Hitchin, and is a registered charity. We hold regular meetings on the fourth Thursday of most months and arrange visits to local buildings and institutions, many of which are not normally open to the general public. We also organize trips to places of historical interest further afield. Members receive a regular newsletter and magazine, the Hitchin Journal. The Society also produces high-quality publications on the history of the town based on research into the origins and development of buildings, organizations, crafts, trades and other aspects of historical interest.

INDEX

A 1927 Advertisement

Stephen Bradford-Best is Hitchin born and bred and has lived in the town for almost all of his life. He retired from a career of more than forty years in the Lord Chancellor's Department of the Civil Service in 2006, having started at Hitchin County Court in Bancroft and finished at the Court of Protection in London. He has an enthusiasm for all forms of transport, particularly vintage cars. He has campaigned his 1936 Austin Nippy sports-car around Europe for the last twenty-five years, covering over 50,000 miles, and has driven *Nippy* to such faraway places as Berlin, southern France, the isle of Sicily and, in 2012, to the Swiss Alps. He is a keen local historian and has written articles for local publications but this is his first book.

'Nippy" in the Alps.(Stephen Bradford-Best)

Stephen at London Zoo.
(Priscilla Douglas)

150